Routledge Revivals

Experimental Analysis of Development

This book, first published in English in 1932, serves as an introduction to experimental embryology. This title, while covering in-depth the field of investigation, presents the general issues surrounding this particular study rather than just providing an analysis of particular results. This title will be of interest to students of introductory biology and the history of science.

Experimental Analysis of Development

Bernhard Dürken

Translated by
H. G. & A. M. Newth

First published in 1932
by George Allen & Unwin Ltd

This edition first published in 2016 by Routledge
2 Park Square, Milton Park, Abingdon, Oxon, OX14 4RN
and by Routledge
711 Third Avenue, New York, NY 10017

Routledge is an imprint of the Taylor & Francis Group, an informa business

© 1932 George Allen & Unwin Ltd

Publisher's Note
The publisher has gone to great lengths to ensure the quality of this reprint but
points out that some imperfections in the original copies may be apparent.

Disclaimer
The publisher has made every effort to trace copyright holders and welcomes
correspondence from those they have been unable to contact.

A Library of Congress record exists under LC control number: 32032529

ISBN 13: 978-1-138-95831-9 (hbk)
ISBN 13: 978-1-315-66123-0 (ebk)

EXPERIMENTAL ANALYSIS
OF DEVELOPMENT

by

BERNHARD DÜRKEN

Translated by

H. G. & A. M. NEWTH

LONDON

GEORGE ALLEN & UNWIN LTD

MUSEUM STREET

First published in German under the title of
"Grundriss der Entwicklungsmechanik"

FIRST PUBLISHED IN ENGLISH IN 1932

PREFACE TO THE GERMAN EDITION

The experimental analysis of development has become an indispensable part of Biology. Very many of those whose attitude towards the fundamental problems of Biology is important are still quite unaware, however, of the problems and conclusions of this branch of study. This applies particularly to medical men. Certain results of outstanding importance have indeed become known, but there is far from being any general comprehension of the real significance of analytical embryology. Often, indeed, we find that the words "experimental embryology" and "analysis of development" convey nothing of the ideas for which they stand. Yet there is perhaps no other branch of Biology which leads the student so directly to fundamental questions concerning the living organism.

This lack of knowledge is partly accounted for by the fact that there is no short book which, while covering the whole field of investigation, presents concisely its great problems rather than its particular results. Considerations of time alone would prevent the readers with whom we are concerned from lightly undertaking the more comprehensive works on the subject.

It seems, too, that there is need of a book of moderate size for those students of Biology who are fortunate enough to be able to study the subject more fully afterwards in the laboratory and the lecture-room; and the need is naturally greater in the case of students who have no such opportunity.

The present book was written to meet these requirements, and it should not be read as a substitute for the more comprehensive works which it cannot replace. It is intended to serve as an introduction only; but as such it may perhaps help to awaken a growing interest in experimental embryology.

In view of the purely didactic purpose of the book, it was considered best to omit from the text all references to literature and original papers. If these are required they will be found in full in my *Lehrbuch der Experimentalzoologie*, and in the other works mentioned at the end of the present book.

B. DÜRKEN

BRESLAU,
July 30, 1929

TRANSLATORS' PREFACE

In preparing an edition of this book for English readers we have not attempted elegance of style. The reader will find—particularly in the more philosophical passages—plain traces of the original German idiom; but he is begged to note that our chief concern throughout has been to convey what we believed to be the thought of the author, without any gloss of interpretation. We have therefore been most literal in those places where the difficulty of the text has offered the greatest temptation to be free.

The mode of employment of certain expressions calls for explanation. The zygote in its early stages of development, before the differentiation of organs, has been throughout called a "germ," as distinct from an embryo—i.e. an Amphibian blastula, gastrula, or neurula would be so called. The convenient abbreviations, "$\frac{1}{2}$-blastomere," "$\frac{1}{4}$-blastomere," "$\frac{1}{8}$-blastomere" have been employed, as is now fairly usual, to indicate a single blastomere of the two-, four-, or eight-cell stages respectively, the denominator of the fraction giving the total number of cells in the germ. The word "regenerate" has been freely used as a substantive to signify that which arises as the result of regenerative processes. If this be a coinage it is justified by its utility and by the precedent of "degenerate."

In accordance with the wish of the author, the word "modification" is always used for the similar German word, as meaning a non-heritable change produced by the environment, though English usage does not draw so clear a line between this and "variation." From the real difficulty presented by *Mehrleistung* we were able to escape only by using a theological equivalent—"supererogation"—which, it is feared, may be less familiar to the younger generation of readers than it is to us.[1]

[1] "Supererogation . . . [f. L. *supererogare*, pay out beyond what is expected], doing of more than duty requires."—*Concise Oxford Dictionary*.

In discussing the relation which the rudiment of an organ may bear to the developed organ itself, Professor Dürken makes use of two distinct antitheses: that between *direkt* and *indirekt*, and that between *unmittelbar* and *mittelbar*. These words he has used in a sense quite special to the context. On consideration we have elected to translate *unmittelbar* and *mittelbar* in the usual way as "direct" and "indirect"; but for *direkt* and *indirekt* the words "immediate" and "mediate" have been used. Since the passages in which these words occur form an important part of the author's thesis, we may perhaps add that "direct" in this usage implies "obligatory," while "indirect" implies "facultative." The words "mediate" and "immediate," on the other hand, imply respectively the mere presence or absence of intermediate steps or processes.

Finally, we wish to thank Professor Dürken for valuable criticism and advice during the course of the translation, and to acknowledge the patience of our publishers in face of many delays.

H. G. N.
A. M. N.

CONTENTS

CHAPTER VIII

THE GERM-CELL AS REACTION-BASIS

ILLUSTRATIONS

EXPERIMENTAL ANALYSIS OF DEVELOPMENT

CHAPTER I

THE PROBLEMS AND METHODS OF EXPERIMENTAL EMBRYOLOGY

I. The Subject of Investigation

1. Classification of Developmental Processes

We have in the processes of development what is by far the most interesting manifestation of life, and what is, at the same time, of the utmost importance for our understanding of the organism. No other function of living things characterizes life so exactly as development. It can safely be said that where there is life there is development also; and that where there is development there is life. For, apart from psychological phenomena, nothing expresses so clearly the contrast between a living organism and a lifeless mechanism as those manifold processes of development, which are seen not only throughout the whole life of the individual, but which pass beyond the individual, linking generation to generation. Developmental processes in their various aspects are, of course, closely interrelated. Hence it is almost impossible to make an entirely satisfactory classification of these processes. The phenomena occur at different times in the life-history of the individual, and are of unequal importance, both from the point of view of the individual and from that of the sequence of generations; it is feasible, nevertheless, to arrange them in such a way as to shew where experiments can profitably be made.

The apparent or *formal* course of development naturally varies with the kind of reproduction, since the starting-points are dissimilar. We shall be primarily concerned with development from an egg, and therefore may make the phenomena connected with this the basis of our survey of the processes of development. Morphogenetic phenomena connected with budding, fission, etc., may easily be brought into the same category.

When we look at the individual phenomena of development, we at once encounter processes which are markedly progressive in character; processes that is to say which, starting from a simpler condition, end in one more complicated and more specialized. Such processes constitute the real content of embryonic development, which begins with the germ in the unicellular condition, and leads up to the final constitution of the organism. But it must be borne in mind that the development of the individual neither really begins nor really ends with that of the embryo. Before the germ-cells have come together in the act of fertilization they have already taken definite steps in development; moreover, after the formation of all the organs of an individual, and after it has become capable of independent existence, processes of development still continue. One cannot, in fact, define exactly—in terms which apply to all cases—the time when these processes cease. We are therefore justified in speaking not only of embryonic development but of pre-embryonic and post-embryonic development, the range and significance of which have limits that are different in different groups of animals. In many cases post-embryonic development includes processes which in other forms still belong to the embryonic period; in other cases, however, post-embryonic processes consist of merely histological changes and of minor structural alterations, which are perhaps only phenomena of growth.

However that may be in particular cases, pre-embryonic, embryonic, and post-embryonic processes can be distinguished as primary and progressive, as against certain other developmental phenomena. These latter have no part themselves in the primary formation of the individual; not only can they appear after the completion of morphogenesis, but in really characteristic cases they never appear earlier. The phenomena to which we refer are those of regeneration. They too are definitely progressive, i.e. they lead from the simpler to the more complex. They must be regarded, however, as secondary processes of development because, normally, they are not intercalated in the course of the primary, embryonic development; and moreover, in their characteristic form, they create substituent structures in place of primary parts which have in some way been lost. It is true that other things than these are included under the term regeneration, as we shall see later; but to these substituent structures we owe the choice of the word, and we shall be content for the time being with this terminology.

Following on the period of primary progressive development, but not sharply divided off from it, there is, for every individual, a longer or shorter period which we can regard as the static period, on account of the small degree of change covered by it. The individual stands at the height of its functional efficiency, and its activities now are what characterize this period—a period which of course varies greatly in length and functional significance in different species. What is important here, however, is that it is during this time that the tissues and organs maintain their normal condition. It may appear illogical to treat from the point of view of development a period whose characteristic is a diminution of developmental activity so great as to make it a state of relative rest; but such treatment is justified when we remember that the maintenance of the tissues and organs already formed is a special function of the organism which can only be reckoned among the developmental functions. In addition, during this static period the true developmental processes of the individual are not all absent. But their progressive character becomes less and less marked—finally to yield to those regressive changes which end in natural death.

During the period of regressive processes there are important transformations. These are not developmental in the sense of being primary and progressive activities of the development of the individual, but retrogressive changes that in no wise lead back to the starting-point of development, but which end by undermining all the normal functions. It need hardly be said that these regressive transformations too do not set in at any definite time, but that progressive ones give place to them by insensible gradations.

Those developmental phenomena which accompany the typical sequence of generations may be said to reach beyond the individual, and can therefore be called " *trans-individual*" phenomena. They may be subdivided into those which are static and those which are progressive. The phenomena which bring about the similarity of developmental processes in all generations—in other words, inheritance—are static. In contradistinction to this we must recognize as progressive that dissimilarity of development in successive generations which forms the main basis of phylogeny. It is, in the main, the static phenomena only which are to be directly apprehended. Progressive processes, on the other hand, can be disclosed only by a logical treatment of the facts, though there are indeed certain matters in this connexion about which we are able to make definite statements, such as mutation, the trans-individual action of environmental factors, etc. Undoubtedly regressive processes too play a part in trans-individual action, and not only do these processes persist in the form of a phylogenetic retrogression of particular organs, but they must also be held in part responsible for the extinction of species or genera.

2. The Experimental Method in the Study of Development

The many individual facts which have been here brought together under the heading of developmental phenomena must be first investigated, on the one hand by direct observation of separate objects, and on the other by the statistical treatment of numerous separate cases. The former method is applicable especially to the development of the individual, the second to the phenomena of inheritance. Thus, there is in most cases a descriptive account of a definite condition, on the basis of which it is possible to build a very convenient descriptive history of development and, up to a point, of genetics also.

But simple observation of the material at hand very soon proves insufficient when a more complete understanding is sought for. Here experimental investigation comes to our assistance in two ways. In the first place it multiplies results by extending the methods of research; but, more important than this, experiment widens the scope of the inquiry itself.

An experiment is carried out when we purposely cause a normal process to occur; so that our account of development can often make itself independent of chance results, and can provide itself with the necessary data without relying on accidental limitations of time and space. But in this aspect experiment is even more useful to genetics, for which statistical facts, systematically arranged, are essential. Modern genetics has only become possible by virtue of the experimental method. The great service of the Augustinian monk, Gregor Mendel[1] was to indicate the way in which it might be effectively applied. Since the rediscovery of his fundamental work in 1900, the exact study of genetics has often been called Mendelism. Experiments in the field of embryology were of course carried out before, even as early as the eighteenth century; but, as applied to processes of individual development, they first became important in the hands of the anatomist, Wilhelm Roux.[2]

In so far as the use of systematic experimentation only involves an extension of the methods of research, its results will remain purely descriptive. For the results obtained shew, at first, simply what happens and how it happens, thus filling up gaps left by the simple observation of nature. No essentially new kind of result is obtained in this way.

With regard to the trans-individual phenomena studied in genetics, the actual experimental results lead up to the statistical (i.e. descriptive) account of the occurrence and distribution of definite products of development (characters) among the whole

[1] Born, 1822, at Heinzendorf, in Austrian Silesia; died, 1884, at Brünn.
[2] Born, 1850, at Jena; died, 1924, at Halle.

of the individuals present. The system of laws deduced from the foregoing relates only to the regular appearance of particular phenomena and to the formal, or descriptive, aspect of their being handed on from generation to generation.

Now, when the experimental method is used not simply to produce intentionally a normal process for our observation, but to change one or more of its conditions, there is then, in addition, a widening of the scope of the problem. The application of this has been mainly in the investigation of separate processes of development, and in the creation of the branch of science called by Roux *Entwicklungsmechanik*. This widening of the problem means that instead of asking what a developmental process is and how it occurs, we now also inquire as to its causes and mode of action; i.e. causal analysis is added to descriptive investigation. That is the real task of *Entwicklungsmechanik*—the "mechanics of development." This term, used by Roux to designate the causal analysis of development, has often given rise to misunderstandings by suggesting the idea of physical mechanics. But the real object of this field of inquiry has no connection with mechanics in the physical sense. Roux, in the choice of this terminology, took as his starting-point Kant's conception of the mechanical event as one which strictly conforms to law, and the word Entwicklungsmechanik only refers to the fact that every event in development is also subject to causality. We are concerned, then, to discover where possible the causes or factors of the processes of development, without, however, adopting in advance any particular theory of the essential nature of these factors. The aim is to disintegrate complicated processes into their simpler constituents by means of experimental analysis. It will readily be understood that researches which attempt to discover the causes and interrelations of development are not always separable from those whose aim is simply an exact account of the course of development. Often it happens that the problem must be presented under both a formal and a causal aspect. Again, mechanical problems—in the physical sense— may sometimes of course be the object of an experiment. We must, however, state emphatically that the real aim of analytical embryology is the discovery of the causes of development without confining the question to the narrow limits of mechanics in the physical sense of the term.

In order to prevent misunderstanding we must distinguish between a mechanical event and—if we may so express it—a *mechanistic* event. In what follows, the former word will always be used in the sense of a "causal event," without the nature of the causes at work being defined and, more especially, without any preconceived idea that only such factors may exist as are amenable to the methods of physics and chemistry. By a mechanistic event, on the other hand, will be meant one in which no modes of action or processes are found other than those in the inorganic world. The first conception is thus the wider of the two, the second being only a part of the first.

The experimental study of development embraces, therefore, a formal and a causal statement of its problems, and its general aim must be to connect all the single results obtained so as to form a coherent picture of what happens in development, and thus to reveal the essential nature of its processes.

II. Special Problems

1. Preliminary

The fundamental problem of experimental embryology can only be solved indirectly by attacking the many individual questions involved. This means the separate investigation of the various groups of developmental phenomena briefly defined above, and of their special constituent processes. If we survey from this point of view what has been accomplished hitherto, we find that these groups have by no means all received the same amount of attention. This is due first to the fact that, for historical reasons, research has had a more or less one-sided bias; second, to the fact that certain phenomena—for reasons concerned with method and technique—have been more easily accessible to experiment.

On this account, in the development of the individual it is principally the progressive processes of true embryonic development and regeneration which have been studied on both the formal and the causal side, while in the case of trans-individual phenomena it has been chiefly the formal aspect of the process of inheritance which has been investigated. It is true that many problems quite different from these have been investigated, but here the results have not been so conclusive as in the groups of phenomena mentioned.

If, beginning with individual development, we consider what are the general problems upon which experiment has been chiefly focused, we shall see that these concern fertilization and the excitation to development. Actual pre-development, important though it be, and though it has been the subject of very thorough study of a descriptive nature, has remained almost untouched by experimental work, for reasons which will be easily understood.

Before the beginning of embryonic development, we encounter processes which, though really belonging to functional physiology, can only be studied in connection with the processes of development: such are the approach of the male reproductive cell to the egg and the penetration of the former into the latter. The activating and determining factors here remain to be ascertained. Pure observation acquaints us with only the externals of the process, experiment alone can disclose what are the forces at work, and what their modes of action.

Closely connected with this is the problem of fertilization. It is a fact of the greatest importance that in bisexual reproduction two different cells constitute the starting-point for the new individual by their fusion to form a single system. The relations of the two cells to one another, and the part they play in the formation of the new organism, are questions of the utmost importance. In the investigation of fertilization and of the activation of the egg to develop, experiments necessarily supplement pure observation, supplying, as they do, descriptive data as well as their own special results from the point of view of causal analysis.

Embryonic development, when looked at from without and superficially, would seem always to begin with cleavage. Descriptively we know many types of cleavage, but that does not suffice to explain the process of cleavage, any more than a descriptive account can explain the cytoplasmic and nuclear movements during the division of the egg into cells. It is for us to demonstrate what forces are at work, and how they are conditioned by the substratum of the cell in bringing about the mass-movements of cells: in other words, our problems are of the order of those of physical kinematics and dynamics. For, whatever may be the primary basis of development, that which occurs during cleavage can be finally reduced to the movements of material parts. The kinematics of the cell is an important subject of investigation, though assuredly it does not cover the whole of the causes of development.

Actual morphogenetic processes begin sometimes during and sometimes after cleavage. To put it briefly, they consist of cell-displacements, of the formation of special layers or groups of cells, of movements and foldings of these layers, and generally, hand in hand with this, of a differentiation of the cells and of the groups or layers of cells. In great measure these processes of morphogenesis can be established by observation alone; often, however, they can only be completely comprehended by the aid of suitable experiments. Our task is—to use the terminology of Physics—to provide a purely descriptive *kinematics* of early development, and at the same time a *dynamics* of the observed movements. Such investigations will exhibit morphogenetic movements in their formal aspect and at the same time the physical forces concerned in them. Great additions will thus be made to our descriptive data, and the mechanistic components of the processes will be established.

It has long been accepted in embryology that when so-called germ-layers are present in the young embryo, each of these layers gives rise to definite organs; but if germ-layers are not formed in the typical manner the rudiments of individual organs can often be traced back to definite cleavage-cells. To put it in general terms: definite parts of the young germ give rise to definite final products of development, which take the form of particular tissues and organs. Each part of the young germ has, in normal development, a perfectly definite fate—or we may say that its presumptive morphological value is strictly defined. In this sense we speak of the prospective value of a cleavage-cell or part of a germ. It is possible to trace back the origin of an organ not merely to the germ-layers or cleavage-cells concerned, but to definite cytoplasmic regions in the unsegmented egg. In certain groups of animals this can be done by observation alone, because visible differences between the cytoplasmic regions exist already in the unsegmented egg. In other forms the presumptive organ-regions can only be determined by experiment—our knowledge of the kinematics of early development has in this way been considerably increased. But that is not all; for the discovery of presumptive organ-regions leads directly to a problem which is indeed analytical. This it is convenient to call the *potency problem*.

2. The Major Problems of Analytical Embryology

That the existence of presumptive organ-regions on the surface of the unsegmented egg can be established has, in the first place, only a purely descriptive, topographical significance; their demarcation shews us simply which parts of the cytoplasm of the egg are later found in particular organ-rudiments; it does not shew that the developmental fate of each individual region is fixed rigidly from the beginning. The development of particular organs is not, in fact, directly due to rudiments, originally present in the uncleaved egg and shewing a definite spatial arrangement: the fate of individual germinal regions is connected only indirectly with the complex of rudiments initially present. Thus it sometimes happens that the fate of a particular presumptive organ-region may be quite different from that which is normal. In other words, its developmental capabilities are not all exhausted in the fulfilment of its normal fate—they embrace something more than what is actually achieved. This is fundamentally true of all germinal regions, though in relatively advanced development it is, sooner or later, modified. This primary developmental capacity which transcends the actual fate of a germ or part of a germ is called its *potency*. The expression prospective potency has also been used, as meaning potency with reference to possible future performance. Our task now is to determine the limits of potency at different periods of development, and to discover how it is modified by the passage of time. This involves not only the potency of the uncleaved egg and its separate territories, but the developmental capacities of blastomeres, organ-rudiments, and those cell-complexes which take part in regenerative processes.

Since the fate of germinal regions is not definitely determined in advance by a rigid mechanism of preformation, there are two phenomena of far-reaching importance which present themselves as being especially important subjects of analytical investigation.

The fact that the potency of a part of the germ may be greater than its normal performance involves the possibility of one presumptive organ-region replacing another without prejudice to the normal formation of a whole structure. This question arises when regions of the germ are lost as the result of experiment, or otherwise. Where such is the case the development of the region that acts as substitute for what is lost, and in certain circumstances the other regions too, must be so *regulated* as to produce a normal whole, in spite of the changed initial situation. Something of this sort often forms an important part of processes of regeneration. The germ and its parts—and, speaking generally, the cell-complexes which are the material basis of development—possess in this way more or less ability to regulate their developmental fate. Though these regulative phenomena may not appear in normal development, it is obvious that the study of them will throw much light upon the special character of the process of development.

Now, if at first there is uncertainty with regard to the developmental fate of the parts of the germ—the characters which they finally assume not being preformed in them from the beginning—then their fate must be decided sometime during embryonic development: their extensive potentialities of action must be narrowed down to what they will actually perform. This is called the *determination* of the parts of the germ. When and how this narrowing-down takes place we learn by investigating potency and processes of regulation, so that problems of this nature are also closely connected with that of determination. The quality of the structures which arise in regeneration must also in some way be "decided," and an investigation of this helps to explain determination, whose course and mode of action enable us to understand how, without an original preformation of the separate organs, a differentiated whole individual can arise.

Determining processes are followed by processes of *realization*, which in their turn bring about differentiation. Determination, strictly speaking, is of course a function of internal developmental factors present in the fertilized egg; processes of realization, however, are influenced in many ways by the environment. The actual form of the final product depends upon two kinds of factors—on internal, inherited factors and on others which exert a modifying influence from without. The environment as a whole, as well as its constituent parts, must be studied on account of its importance in development. In this connexion the much debated question arises as to whether the environment influences only the individual, producing modifications only in individuals, or whether it has also a *trans-individual* action; i.e., whether the action of the external factors appears only in the generation immediately affected, or is carried over to later generations. This problem is often discussed in connexion with phylogeny, but it also has a significance and importance quite apart from this.

Determination of a part of the germ never occurs in such a way as that its fate is decided apart from the other events of development. On the contrary, it always has reference to the whole, and is in connexion with the whole. Thus, development does not simply consist of a sum of independent processes running side by side, but is one undivided activity in which all the processes and all the parts of the germ are closely knit. These interrelations during development constitute its most essential characteristics. We find determining, realizing, and modifying interrelations, together with others, in both the primary and secondary processes of development. From the purely descriptive point of view these interrelations appear in the form of a great variety of both reciprocal and non-reciprocal kinds of interdependence. Liberation of the parts from such interdependence can be seen to occur after determination is complete, so that there is in the end a certain mutual independence of the parts. But apart from the mere existence of such interrelations, our concern is with the agencies through which they are made effective; and the questions then to be decided are whether, and to what extent, these agencies are material and chemical in nature or are purely energy-relations; how far direct mechanistic action is at work, as against that of stimuli which produce physiological reactions; and so forth.

Investigations of this kind into the individual processes of development have resulted in a particular view of its essential nature and of the factors concerned in it. Now, such conceptions are also formed on the basis of the phenomena of trans-individual processes; genetic research, independent of the analysis of development, has, indeed, formed conceptions and adopted points of view relating to the nature of development and of its factors. Further, by collating these ideas with the results of cytological research, genetics has been able to consolidate the basis of its conceptions. In comparing the fundamental ideas of the embryologist with those of the geneticist we discover a striking contradiction. Since, however, both the analysis of development and genetics are really concerned with the same thing, namely with internal factors of development that are neither more nor less than inherited bases of development, a unified conception of the nature of these factors and of development itself should be possible. We are thus confronted with the task of reconciling the discrepancies between the fundamental ideas of two fields of research both concerned with development. In attempting this, however, we must pay special attention to the relative competence of the two methods of research to deal with particular problems.

At that point we will leave out survey of the field of analytical embryology. Naturally it has been possible to give only the main outlines of the subject: the major problems indicated can only be grasped by dealing singly with the lesser problems they include. Given, however, a knowledge of its separate processes, we are still far from understanding development as a whole, if our aim is the characterization of the living organism. The organism is always a unit and a whole, and not simply a sum of separate processes. From this fact again arises a multitude of problems, whose solution must depend upon the study of development as a whole.

III. Methods of Analysis

1. General Observations on Method

The method of experimental embryology is in principle analytical. It consists in a dissection of complicated processes into their simpler parts, and in a dissociation of factor-complexes, themselves not clear and comprehensible, into their constituent factors. In such analysis the causal point of view often retreats into the background; for it is first necessary to determine merely what constituent processes are present; in the second place our concern is with the mode of interaction of these constituents and their union to form a developmental unity. Thus it comes about that, in applying this method, we find that in practice we cannot strictly separate descriptive from causal investigation.

Analysis, of course, is not an end in itself, and therefore cannot be the final goal of research. The function of analysis, apart from furnishing an exact knowledge of the separate parts, is to provide the elements for later synthetic treatment. Descriptive morphology has also become an analytical science, dividing the organism as it does into ultimate morphological constituents. We are not, however, justified in claiming that a knowledge of the ultimate morphological elements enables us to understand completely the structure of an organism. Despite the fact that the organism is made up of morphological constituents—ultimately the cells and their proper parts—it remains a whole—an individual. Even from the morphological standpoint a true estimate of the organism can only be arrived at by bearing in mind its wholeness and unity, i.e. by means of a synthetic conception. As a result of the classical Cell Theory, this point of view has become unpopular. There is now, however, a widespread reaction against the over-valuation of the individual cell and its constituents, and the organism as a whole begins to take a prominent place, but without an under-valuation of analytical results being thereby involved. If the relation to the whole organism is lost sight of, the result of analysis remains a mere collection of meaningless fragments. The importance of the whole in developmental processes is abundantly verified by the results of analytical embryology.

A point we shall recur to later is the dependence of the nature of the result upon the method used. Physical methods yield only physical results; chemical methods, only chemical results. Increase of our knowledge is thus limited by the scope and nature of the methods at our disposal. We are therefore not justified in excluding *a priori* the possibility of results other than those obtained by the methods in use, since naturally other methods may be available in the future. In experiments on an organism by physical and chemical methods—that is by the use of physical and chemical agents—no phenomena will be established except those which can be comprehended mechanistically; but it does not follow from this that the living organism can exhibit no other kinds of phenomena and of processes. We need only use a method that is, as it were, really biological or *organismic*, in order to obtain specifically organismic and really biological results. Developmental analysis to-day fortunately provides such an extension of method, if at present only to a limited degree. This is illustrated by cases where, for example, the mode of action of a living developmental complex, instead of being chemically or physically

c

analysed, is tested with regard to its ability to perform some specific process in development; or where this mode of action is investigated by allowing two living systems to influence one another—adopting as criterion of their action some morphogenetic event in the living organism. It is the special character of the results so obtained that makes the analysis of development invaluable for a just comprehension of the nature of the organism.

In all phenomena of development two groups of factors are concerned: internal factors, and external or environmental factors. To discover individual factors and their modes of action is the first aim of experimental analysis. We may roughly distinguish two main types of experiments: procedures depending on addition, and procedures depending on subtraction.

A particular complex of external factors is necessary for normal development. It is possible to break up this complex in two ways: (a) by removal of a particular component factor from the environment (subtraction procedure), or (b) by adding a new component factor to the normal environment (addition procedure). For example, in the one case the action of light may be entirely eliminated, while in another case special kinds of rays—e.g. of short wave-length—may be added to normal daylight. As applied to internal factors the principal subtraction procedure is that which is involved in so-called "defect experiments." Certain definite cells, or rudiments of organs, can be removed from a germ and the fate in development of the rest can then be followed. As addition experiments in connexion with internal factors we may cite the artificial joining of one germ to another germ, the former thus being brought into the sphere of action of the latter. Crossing experiments are a means of studying the trans-individual behaviour of factors by an addition-technique.

The detailed arrangement of an experiment depends entirely upon the particular problem attacked, and often involves a very difficult technique, an account of which would be out of place here. Something must be said, however, about the general principles of a few experimental methods which are based on phenomena that have no place in ordinary development. Such phenomena are those of regeneration, transplantation, explantation, and interplantation.

2. The Principles of Certain Specific Procedures

(a) *Regeneration*

The production of secondary, regenerative processes of development provides an experimental method which is much used. The processes in question are not only interesting in themselves, but the behaviour of regenerated parts adds much to our knowledge of potency and regulation, and of other matters concerning the internal factors of development. The ability to form "regenerates" is widespread. It must not be regarded as an adaptation to frequent loss of organs or parts of the body, but as a primary characteristic of the organism. This is shewn by the fact that power of regeneration and frequency of loss of parts by no means go hand in hand. It must also be noticed that in certain circumstances structures are "regenerated" without a loss having previously taken place; there are, indeed, cases where the regenerate does not arise in the place where the injury occurred, but in an undamaged area. Therefore we should, to be exact, designate the regenerate not as a substituted structure but as a product of secondary developmental processes which the organism can carry out even after the completion of its true embryonic development.

It may be said that in general the power of regeneration is inversely proportional to the degree of organization and the stage of development of the individual. Thus the so-called lower animals show this power in the most marked degree, while in the Mammals, for example, it is only slightly in evidence. It is also slight in Birds, while in Amphibians and Fishes it is considerable. Ability to regenerate is greatest in invertebrates, though here too there are, naturally, different degrees. In no group of animals, however, is the ability to produce regenerates altogether lacking. With age this power decreases: in all organisms it is more noticeable during the period of primary morphogenesis, after which it is, in general, obviously reduced. Young frog larvæ, for example, can still to some extent regenerate whole limbs normally, but in metamorphosed frogs this never happens, though it does in adult Urodeles. It is hardly necessary to emphasize the fact that the power of regeneration and the power of multiplication by fission and budding are closely related.

In regeneration the structure is not simply formed from cell-material already present, but a so-called *blastema* first arises from more or less indifferent cells. In this the differentiations concerned very often appear exactly as they do in embryonic development. These processes vary, of course, in details, but generally a temporary closure of the wound first takes place; the new blastema—in which active cell-proliferation takes place—is then formed from cells generally derived from around the wound-surface; very often the formation of the blastema is preceded by processes of shrinkage near the wound. In certain circumstances the blastema forms a conical projection, and may be called a *regeneration-cone*. This is not to say that the new formation always arises from what still remains of the tissue or organ removed; on the contrary, it is often unnecessary that there should be any such remainder. The blastema is not always so extensive as we frequently find it in Vertebrates. In Arthropods, for example, it is only represented by the slightly thickened hypodermis which forms again under the wound.

In invertebrates, besides a new formation of tissues, there occurs a rearrangement and transformation of cellular material already present; and in some cases the normal condition is re-established solely by such "regulation."

The origin of the cells in the true regeneration blastema has not yet, in all cases, been satisfactorily explained. In worms, for example, large amœboid cells are concerned; these migrate from all parts of the body to the site of the injury; but in some animals tissue-cells which are already differentiated are responsible for these new formations—a phenomenon of great general importance.

Repeated regeneration of one and the same organ may, in certain circumstances, occur: the regenerated parts themselves possess a power of regeneration, so that we might speak of regenerating regenerates. Again, it is not only appendages and external organs, but also internal and protected parts that can regenerate, so that this capacity must be ascribed to all living tissues as fundamental.

A very remarkable phenomenon to be noticed in connection with regeneration is self-mutilation, or *autotomy*, which is met with in the most widely different forms of animals. This is the

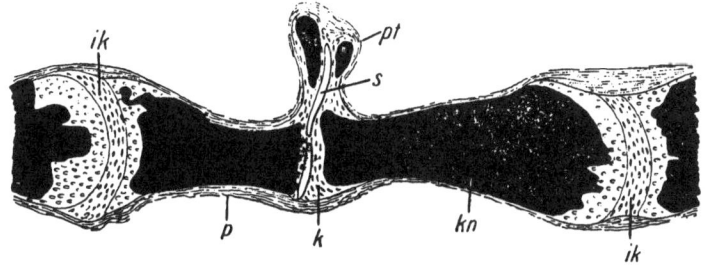

FIG. 1.—Arrangement of transverse breakage-planes preformed in the tail vertebra of the lizard; the autotomy-fissure traverses also the vertebral processes. *pt* transverse process of the right side (the section has missed that of the left side); *s* split; *k* cartilage of the cleavage-plane; *ik* intervertebral cartilage; *kn* bone; *p* periosteum. (After Slotopolsky, diagrammatized.)

spontaneous casting-off of single parts or organs, usually for no apparent reason, but sometimes as a reaction to strong stimulation. As a rule autotomy takes place as the result of strong muscular action, and either at any point in the organ, or at some place specially prepared. In the latter case breakage joints are present, especially in Arthropods. Here they generally consist of an annular thinning of the chitin covering the limbs, which makes possible the active or passive breaking off of appendages. In worms, Echinoderms (starfish), and others, there are no special breakage places, but parts of the body are constricted off at any point by strong muscular contraction. Among Vertebrates, lizards possess special arrangements for autotomy (Fig. 1). Preformed breakage places are found in the caudal vertebræ, usually from the sixth onwards. They consist of transverse fissures, which do not pass completely through the vertebræ but leave, here and there, bridges of cartilage. These last hold the two halves of the vertebræ together, but at the same time allow it to be broken across. Corresponding to this preformed breakage-surface the connective tissue of the tail is traversed by two closely apposed tendinous plates. These reach as far as the integument, so that connective tissue and muscle masses are also separated into two halves. This arrangement for casting off the tail works as a reflex mechanism, which is not controlled by the brain. The lizard thus loses its tail not by violent mechanical rupture, but by means of an autotomy mechanism activated by direct stimulation.

Regeneration of the lizard's tail occurs not only from preformed breakage places, but also from other places at which it has been amputated.

(b) Transplantation, explantation, and interplantation

In the analytical investigation of development it is often necessary to isolate individual parts of the germ, or individual tissues and organs, by removing them from their normal position, and thus from the action of factors to which they are exposed in the complete organism. Further, part of one germ may be brought into the sphere of action of another germ, and thus the interaction of two living complexes be studied. In making such experiments the procedure is that of transplantation, explantation, or interplantation, according to the particular end in view.

By transplantation, or grafting, is meant either the artificial union of two organisms, or the transference of some part or organ to a different position in the same individual, or to any position in another individual. Apart from the purely scientific importance of transplantation, it has great practical value in surgery.

We have pointed out above that there are different kinds of transplantation; these are usually distinguished as autoplastic, homoioplastic, heteroplastic, and xenoplastic (Fig. 2). When a transplant or graft taken from a particular individual is reinserted in a new position in the same individual we speak of auto-plastic transplantation; the transplant generally fuses well with the host. In homoioplastic transplantation a specifically similar transplant is involved; that is, the transplant is taken from an animal of the same species as that to which the host belongs—the experiment is confined to a single species. Such transplants also may establish themselves well and possess a good viability. In heteroplastic transplantation, transplant and host belong to different species or even to different genera. Usually in such experiments the graft shews only a limited capacity for establishing itself; in the most favourable cases it suffers pronounced degeneration, but generally it perishes sooner or later or is cast off. Xenoplastic transplantation is allied to heteroplastic and plays a somewhat important part in experimental embryology. It is a transplantation or exchange of parts between individuals of different species or genera, but, in particular, the introduction into a young germ of parts of another germ which is specifically or generically different. The essential difference between heteroplastic and xenoplastic transplantation is that in the latter case graft and stock stand in the relation of guest to host—i.e., the graft remains alive a long time and may even establish itself completely in its new surroundings. Its further fate in development may accord either with its place of origin or with its actual, secondary position.

The practicability of transplantation in organisms is closely connected with their power of regeneration: it corresponds well, both as regards its distribution and its degree, to the power of regeneration possessed by their component parts. Trans-plantations are possible in all animals—from the lowest orga-nisms to man himself—and at any age, in the germ no less than in the adult organism. The parts that can be successfully transplanted are extraordinarily manifold; there is hardly an organ which cannot be transplanted, though in special cases there may be limits, due to the special nature of the organism or to technical difficulties. The practicability of transplantation diminishes with the age of the organism and with increase in its degree of organization.

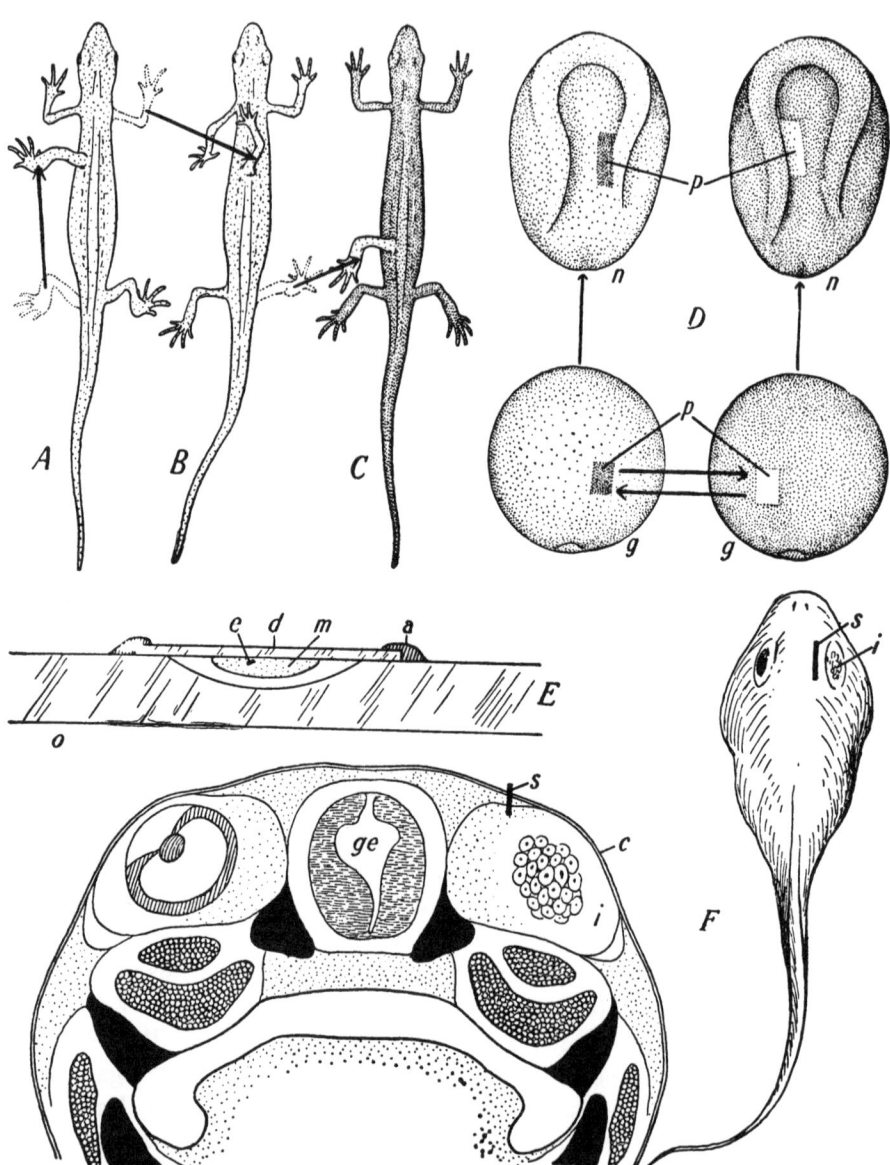

Fig. 2.—Diagrams explaining transplantation, explantation, and interplantation. *A–D*, transplantation. *A* autoplastic (transplantation in one and the same individual); *B* homoioplastic (transplantation between two individuals of the same species); *C* heteroplastic (exchange between two individuals of different species or genera—not very stable); *D* xenoplastic (exchange between germs belonging to different species or genera; relation of implant to recipient, showing good viability of the former); *E* explantation (planting-out into a nutrient medium); *F* interplantation (insertion into the cavity of an organism, which serves as culture-chamber); *p* implant; *g* gastrula; *n* neurula with medullary folds; *o* microscope slide; *d* cover-glass; *a* wax or vaseline seal; *m* nutrient medium; *e* explant; *i* interplant; *c* conjunctiva; *ge* brain; *s* line of incision.

The older and more complicated the transplant, the more does it tend to show early phenomena of degeneration, which are most inimical when what is wanted is a permanent transplant capable of functioning—as in surgical practice. All that is often wanted in experimental embryology, however, is a temporary union. In a highly organized transplant degenerations appear even if the operation has been auto- or homoioplastic. Thus, a transplanted bone always itself perishes, its periosteum alone may remain and replace the resorbed bone by a new formation. The experiments so far made seem to shew that heteroplastic transplants, in the case of adults of the more highly organized animals, always, in the end, undergo complete degeneration.

The ability to maintain itself which is possessed by a transplant depends on the amount of specific or individual difference between graft and host. It appears, in fact, that the ease with which transplantations can be made shews a gradation, such that transplantation is not merely more practicable between individuals of the same species (homoioplastic) than between those of different species (heteroplastic), but is more practicable between "blood-relations" than between individuals which merely belong to the same species.

Every species is distinguished from others by specific differences; and in the same way within the species each individual is distinguished from the others by individual differences. The question of what these ultimately are need not concern us here; undoubtedly peculiarities of the physiological chemistry of the individual are involved. After implantation of a complex part the tissues of the host shew definite reactions. Leucocytes and lymphocytes accumulate in the neighbourhood of the graft, and together bring about the processes of resorption which sooner or later appear. The greater the specific difference, the earlier and the more marked is that accumulation. The inimical behaviour of the host towards the graft— as towards a foreign body—is the chief explanation of the lesser stability of heteroplastic in comparison with homoioplastic—or indeed autoplastic—transplants. Comparable—though naturally weaker—reactions occur when host and graft are derived merely from different individuals of the same species. This even holds for the closest relationship of donor and recipient. Among blood-relations the defensive reactions (e.g. in tissue transplantations in guinea-pigs) are strongest when the graft is from child to mother, weakest as between brother and sister, while when from mother to offspring they appear to be intermediate in intensity.

These actions of individual and specific differences have an importance which is not purely theoretical. They affect the practice of surgery, since they endanger the stability of homoioplastic transplants—to say nothing of heteroplastic transplants.

When pieces of tissue, organs, or parts of organs, etc., are removed from an organism and, instead of being reimplanted in an organism, are brought into a special medium in which the parts in question can continue to live, we call this *explantation*. Other names for this procedure are tissue-culture, coverglass culture, and culture *in vitro*.

The choice of a nutrient medium is here naturally of great importance. The most suitable has proved to be the blood plasma of the same animal, in certain circumstances with an admixture of embryo-extract of the same species. Blood serum, lymph, and even artificial solutions have also been used with good results. In general the procedure is to cover a particle of tissue on a cover-glass with a drop of plasma, and to place the cover-glass, object downwards, on a slide with a cavity ground in it. The edge of the cover-glass is then sealed in some way so as to prevent evaporation or other damage, and the culture maintained at a suitable temperature, preferable bodytemperature, in an incubator.

Phenomena of growth and cell-multiplication are chiefly observed at the edge of the explanted tissue. In this way there is generally soon formed round the explant a halo of cells which have wandered out and which have newly arisen by division. The growth is chaotic—that is, multiplication of cells is diffuse, without the differentiation of specific tissues and, when the medium is homogeneous, with no preference for any particular direction. Cell movements and wanderings play an important part; these, however, only occur when the cells are in some way subjected to contact-stimuli (thigmotaxis). Wanderings of epithelial cells within the medium occur in plasma cultures, but not in serum cultures. It is possible that, in addition to chemotaxis, we are here concerned with a certain amount of minute structure in the medium, caused by tensions within it.

If the cultures remain alive and it is desired to continue their growth, the younger parts of the explant must be transferred from time to time to fresh nutrient medium. In this way it is possible to keep tissue cultures alive for many years; and the individual from which the original culture was made may have been dead for a long time while explants from its tissues continue quite fresh and viable. The fact that tissues may, if the conditions are suitable, be maintained outside the body without their continued life and growth being adversely affected, is a result of great general significance.

Hitherto tissue cultures have been used principally in the study of tumour cells; and only latterly has the procedure begun to assume any importance in connexion with problems of experimental embryology—in fact since it was established that the nature of the growth and the differentiation that occur depend upon the character of the culture medium.

Finally we must consider another procedure which may be called appropriately *interplantation*. In the case of true transplantation the transplant is cut off from its normal relations, it is true; but it is brought into new relations, since it actually fuses with the new environment. Frequently, however, it may be desirable to bring about a more complete isolation of part of a germ, in order to investigate the fate in development of such a single complex, freed as far as possible from external influences. This can be done by removing parts from a young germ and placing them in a cavity of the body of an older animal. Generally such an interplant remains, throughout, independent of the host. It behaves as a foreign body, without entering into any real connexion with the tissues of the host, even when it develops and differentiates. Though in some cases the interplant may be resorbed, it generally persists and thrives, so that the purpose of the experiment—to observe its behaviour over a considerable time—is well served. Various cavities have been employed as places for interplantation, e.g. the coelom, and the perilymph cavity of the labyrinth. Specially suitable, at least in the Amphibia, is the cavity of the orbit (Fig. 2, F). For example, the eyeball of the young frog-larva can be removed by making a cut dorsally above the eye, taking care not to damage the already transparent corneal epithelium (conjunctiva). The empty orbital cavity then provides a culture chamber covered over by the epithelium as by a watchglass. With a suitable pipette pieces of the germ, such as particular parts of the blastula or gastrula, the developmental fate of which is to be tested, are introduced into the chamber through the cut, which quickly heals.

The isolated cell-complexes require no further attention, since they are protected against injury, and sometimes catabolic products are removed with the excretions of the host. So long as the corneal epithelium remains transparent the culture, even in the living host, can be observed from outside. In this way interplanted germinal regions may develop to form highly differentiated tissues or parts of organs. Even though every influence on the part of the host cannot be entirely excluded from the beginning, the isolation of the germinal regions so treated is at least very considerable; in particular they are entirely removed from the moulding influences of the rest of the germ. As a rule the interplant is, where possible, introduced into a host of the same species, to avoid unwanted phenomena due to specific differences. It is not, however, absolutely necessary to keep within one species. For example, young larvæ of *Salamandra maculosa* have proved to be suitable hosts for parts of specifically and generically different blastulæ and gastrulæ; even parts of the germs of Anurans have been successfully reared in their orbital cavity. Such interplantations occupy an intermediate position between true transplantations, where actual fusion occurs with the host, and explantations, in which the culture is upon a non-living substratum.

CHAPTER II

FERTILIZATION AND THE STIMULUS TO DEVELOPMENT

I. INSEMINATION

In bisexual reproduction the development of the new individual begins with the fusion of the egg and the spermatozoon. This fusion at first sight seems to be a perfectly simple process. Sperms set free by the male within a certain distance of the egg (which remains motionless) seek this, aided by their active movements; by the penetration of the male cell into the female the fusion of the two gametes is completed; and, in general, embryonic development begins at this moment. Closer observation shows that insemination is by no means such a simple process, but that a whole series of separate factors take part in it. The circumstance that we are not limited to chance observations in the study of insemination, but can bring about the process at any desired moment artificially and even interfere with it experimentally, makes it possible to investigate the separate individual factors and their mode of action.

Artificial insemination (which is by no means the same as artificial fertilization) is possible in all forms of animals. For experimental purposes we are concerned only with animals whose eggs and sperm come together outside the body—very largely, therefore, with Fishes, Amphibia, Echinoderms, etc. Eggs taken from a ripe female are simply mixed with seminal fluid taken from the male, or brought into contact with a suspension of sperm or particles of testis in water. Apart from its scientific aspect, artificial insemination in certain circumstances is of great importance in the practice of animal breeding, e.g. in pisciculture and occasionally even in the breeding of mammals. Many hybridizations can only be made with its help.

The incompleteness of the study of insemination makes it impossible to analyse the process in all its details; neither can one say with certainty that results based on a particular species are of general application. With these reservations made, however, the following survey can be given (cf. diagrammatic survey in Fig. 3).

The process of insemination is led up to by the discharge of the gametes from the gonads and, in the case of the spermatozoa, by their emission from the body. The eggs remain in many cases within the body of the female, but often they are separated from it before insemination. In the former case a specific copulation takes place; in the latter spermatozoa are emitted in the neighbourhood of the eggs. From the point of view of insemination this is not a fundamental distinction; for even in the case of copulation the egg is not brought into immediate contact with the sperm and, whatever the circumstances, the latter has to travel a certain longer or shorter distance in order to reach the egg.

It is primarily the random movement of the sperms themselves which brings about their approach to the egg; and many sperms fail in this chance fashion to reach their goal. This applies especially to the case in which both kinds of gametes are simply shed into the surrounding water. But approach to the egg is aided by the chemotactic susceptibility of the spermatozoa. Where there is intromission this chemotaxis begins to be of importance at some distance from the egg, since the secretions of the female organs, at least in part, exercise a positive chemotactic attraction on spermatozoa and thus aid further penetration. Where unfertilized eggs are laid in water the chemotaxis of the male cells first makes itself felt only when they are much closer to the eggs. In these cases it must be substances set free from the egg, or alterations by the egg of the concentration-gradient of substances in the water, which bring about the attraction of the sperms (Fig. 3, a). Whether something of the sort takes place in all cases, even when the egg remains in the female oviduct, remains to be proved. Such an attraction of the spermatozoa by substances emitted by the egg, and by consequent changes in the concentration-gradient of the environment, is found in all Echinoderms.

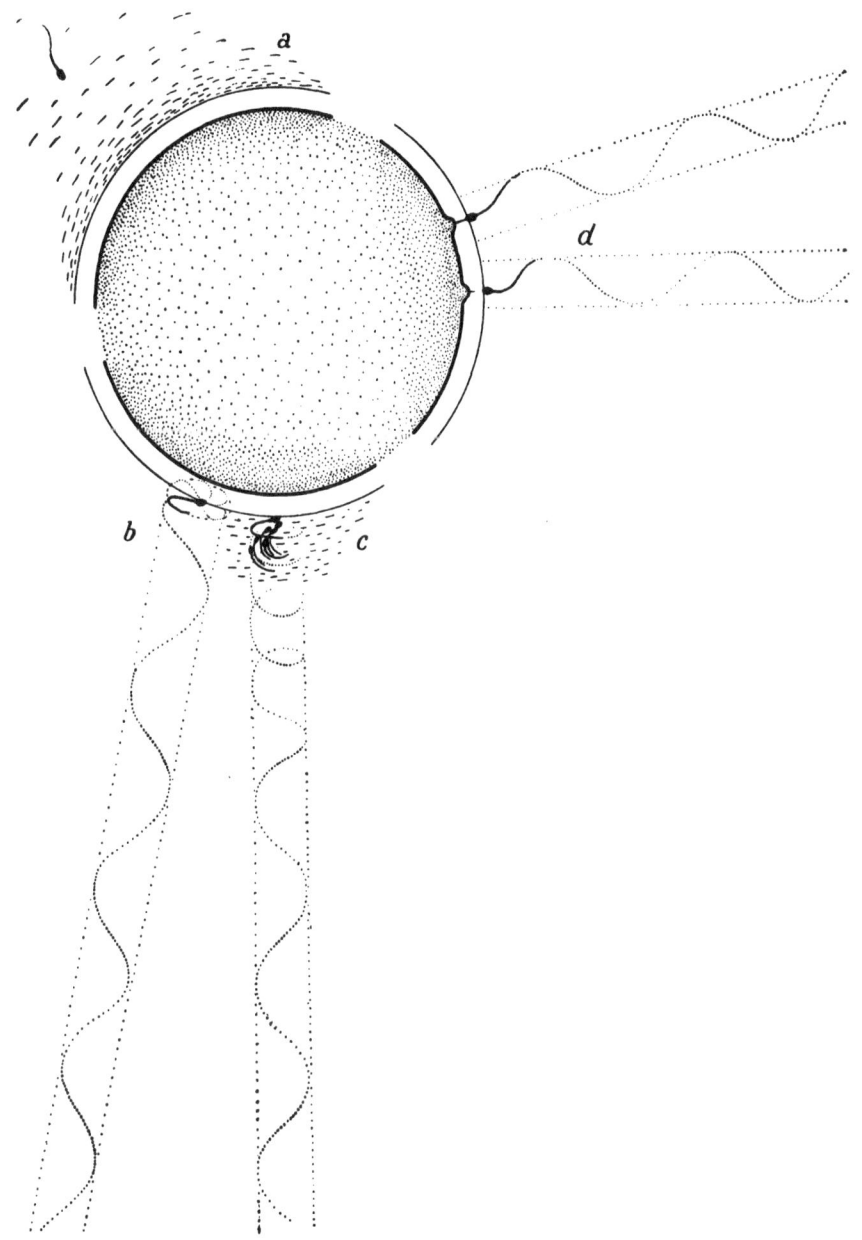

FIG. 3.—Diagrammatic survey of the various processes taking part in insemination (explanation in the text).

Though the substances set free by the egg may act in such cases as stimuli to spermatozoa, and as a means of attracting them, increase of motility is also a means, on the other hand, of keeping foreign spermatozoa at a distance and of favouring monospermy. Spermatozoa, when freely moving, follow a spiral course, even though the spiral may be so steep that it is hardly, or not at all, perceptible. Those sperms—relatively speaking—most easily penetrate the egg-envelope which impinge upon it when approaching in a radial direction. Now, if the spermatozoa come into contact with a surface, e.g. with an air-bubble or with the egg itself, the turning movement quickens under the influence of the contact-stimulus, so that each spiral ends by flattening out, and the spermatozoa carry out serpentine movements tangential to the surface. Penetration is thus prevented (Fig. 3, b). A similar tangential position is also seen under influences which are clearly chemical, when e.g. sea-urchin eggs and starfish sperm are brought together. In the normal fertilization of sea-urchin eggs by their own sperm one also sees such tangential, serpentine movements, which as we have already said are chiefly aroused by contact stimuli. Since, however, all the sperms do not take up a tangential position, but many are radially disposed and their penetration is thus made possible, one must in the latter case assume a lessening of excitation, due perhaps to substances emitted from the egg, We here distinguish then on the one hand a factor favouring specific fertilization, on the other hand one favouring monospermy.

A further protection against the penetration of foreign spermatozoa is provided by agglutinating substances, present in the gelatinous membranes of eggs, which cause foreign sperms, and in some degree also sperms of the same species, to adhere together (Fig. 3, c). In Echinoderms, again, a specific poison with a lethal action on foreign sperms diffuses from the body tissues, or from specific glands, into the surrounding water. We cannot at present determine how far these phenomena are present when the eggs are fertilized within the female body; the possibility must be admitted that there too they play a part.

All this only succeeds in bringing the spermatozoa into the immediate vicinity of the egg. In the actual penetration of the spermatozoon other processes are concerned, which—at any rate at present—cannot be referred to immediate causes. When individual spermatozoa have penetrated with their heads into the gelatinous envelope of the Echinoderm egg, they come to rest. The sperm head does not bore into the substance of the egg, but an active participation of the egg now occurs (Fig. 3, *d*). On its surface there are formed conical protuberances which have been called entrance cones, and each of these sends out a fine process or filament through the jelly to the motionless sperm head. That filament which first seizes a spermatozoon draws back, pulling the spermatozoon into the entrance cone; at the same time all the other filaments are drawn in, whether they have seized a spermatozoon or not. Arising from the cytoplasmic cone whose filament first reached a spermatozoon, a characteristic fertilization-membrane is formed, and this forms a barrier against the entrance of further spermatozoa. As we have pointed out, it has not been explained how the activity of the egg cytoplasm is evoked; it is at first sight suggestive of action at a distance. The apparently simple process of insemination thus covers a number of important problems.

II. The Factors which Excite Development

1. Artificial Parthenogenesis

Generally the beginning of embryonic development follows immediately on the insemination of the egg, as is shown by the beginning of cleavage. The first question to arise is whether the essential function of fertilization is solely that of provoking development: whether, that is to say, insemination and fertilization are the same thing; or whether we should regard them as distinct though generally concomitant processes.

Now the facts of natural parthenogenesis, in those cases in which the egg develops as easily with as without fertilization (e.g. the honey-bee), shew that fertilization and the stimulus to development are not necessarily connected. Certain conditions, might, however, easily escape observation alone, so that wherever possible an experimental analysis must be attempted. That is by no means difficult; eggs which normally require fertilization can successfully be caused to cleave, and even to develop, without it.

If, on purely objective, morphological grounds, the completion of fertilization be regarded as the union of the paternal and the maternal nuclear substances present in the male and female pronuclei, then it is this union which must be prevented without causing cleavage or development to cease. This end can be most simply achieved by destroying the male pronucleus by irradiation with radium.

The β and γ rays of radium affect chiefly nuclear substances; these substances are destroyed, while—with correct dosage— the rest of the cell materials remain undamaged. If spermatozoa are suitably irradiated with radium, the nuclear substances lose their power to grow and to multiply, though the sperms retain their motility and their ability to penetrate the egg. Further investigation shews that no normal pronucleus is then developed from the spermatozoon. The sperm nucleus remains unaltered in the cytoplasm of the egg, passes during cleavage (which nevertheless is quite normal) into one or other of the blastomeres, and finally completely degenerates. If for example the sperm of *Triton* is irradiated during $2-2\frac{1}{2}$ hours with a strong preparation of meso-thorium, which has the same action as pure radium, and if then normal eggs are inseminated with this sperm, there are developed larvæ which may live for several weeks. Development proceeds on the whole normally, but the larvæ are noticeably smaller than those from fertilized eggs, and ultimately shew various defects. Thus, the gill filaments are only poorly developed; the gelatinous connective tissue tends to show hypertrophy; and among other malformations there appear embryonic proliferations in the brain, the spinal cord, and the retina.

Now what is really important here is that these larvæ have a parthenogenetic origin, that is they contain no nuclear material of paternal origin: no fertilization has preceded development. This can be assumed from the above-mentioned behaviour of the sperm nuclei when damaged by radium rays; but it can also be proved by counting the chromosomes in the advanced larvæ. From the male and female pronuclei at fertilization there arise, as is well known, chromosomes in equal numbers, which together form the basis of the zygote nucleus, and therefore of the nuclei of the cells of the new organism. The ripe germ-cells contain only half the number of chromosomes (haploid number) which is characteristic of the body cells; in fertilization the normal, or diploid number is again established. The diploid chromosome number for *Triton* is 24. If then these larvæ have indeed arisen without fertilization, solely as a result of insemination—in other words if they are

Fig. 4.—Chromosomes in the mitosis of an epidermal cell from the fin of a parthenogenetic newt-larva 24 days old.

parthenogenetic in character—then one would expect to find only 12 chromosomes in the nuclear division of their body cells. This is actually the case (Fig. 4). The count is best made in the nuclear divisions of epidermis cells in the median fin of young larvæ, where 12 chromosomes can be seen. To this smaller number of chromosomes there corresponds again a lesser size of the resting nuclei as compared with their diameter in normal larvæ from fertilized eggs. Experiments on other kinds of animals have fully confirmed these statements.

It is only necessary to add here that, in certain cases, the haploid chromosome number may be abolished by subsequent processes of regulation. That fertilization is separable from the stimulus to development is thus established, and it is proved at the same time that fertilization *sensu stricto* is something more than the starting of the developmental processes.

Activation is not, moreover, a specific vital function of the spermatozoon, but a process which can be brought about by the most diverse influences, and—what is most important for an understanding of it—by purely mechanical means. As a general principle we may say that every egg, of whatever kind of animal, can be stimulated to cleavage not only without fertilization but without insemination, and in this way a normal individual can arise from a single germ-cell. It is a matter of indifference whether the egg in question normally needs fertilization or not. The ability to develop must be ascribed to every fully formed egg as a fundamental property. A stimulus of some kind is required in order to set in motion the processes of cleavage and development. From the point of view of experiment we must consider the following: mechanical shaking; puncture of the egg with a fine needle; a temporary (especially a sudden) rise of temperature; the effect of hyper- and hypotonic solutions, and in general of substances which change the surface-tension of the egg in relation to the surrounding medium. Chemical agents too of the most diverse kinds, such as salts, acids, etc., can in the right concentration provoke cleavage and development. The success of one of these artificial influences rather than of another depends upon the kind of egg used; but in principle we may say that every egg can be made to develop without the help of the spermatozoon.

In the case of the frog's egg the best method is that of pricking it by the slight jerking movement of a glass needle about 10μ to 15μ thick. It is not necessary to choose any particular pole of the egg for the puncture. Eggs taken out of the uterus are smeared in a layer on a slide, pricked, and then put in shallow water. The eggs which have been pricked behave exactly like normally inseminated ones. After the extrusion of fluid the egg becomes able to turn round in its envelopes, and at once places itself normally—that is with the heavier, vegetative pole downwards. The first sign of development is a flattening at the dark pole—which is also characteristic of the normally inseminated egg. The so-called grey crescent—to which we shall refer later—appears. As a result of the prick, more or less of the substance of the egg flows out of the puncture; this *extrovate*, which persists for a long time, is pointed when a thin needle has been used, and mushroom-shaped when the needle was thicker. The former are better. The first cleavage usually appears at the same time as it would do after insemination. The course of cleavage is in many ways irregular; in many eggs the flattening which we have mentioned is the only sign of the reaction. Only comparatively few eggs get beyond gastrulation, but a few at least will develop normally further than this, so that metamorphosed frogs may be obtained. Sometimes these are dwarfed, but they may be indistinguishable from normal frogs.

Among the chemical agents which are capable of setting in motion parthenogenetic cleavage and development are the salts of potassium, sodium, calcium, and magnesium: especially good results are obtained with sea-urchin eggs when $MgCl_2$ is used. It is not necessary that the substance used should be an electrolyte: development is also started by the addition of sugar solution.

Rise of temperature is most successful in producing cleavage if it is allowed to act before the formation of the second polar body. The degree of temperature employed and the length of time necessary vary with the kind of egg.

2. Analysis of the Process of Cleavage

It will be readily understood that other factors than those present in natural insemination may be at work in the case of the activation occurring in artificial parthenogenesis. This difference only applies to the activating factor which first impinges on the egg. For, since cell division is normal in both cases, the egg must be placed by the different factors in such a position as to produce its cleavage. Since there is already present in the egg a general capacity to divide, the similar effects of different influences can only be explained by supposing that the specific nature of the cytoplasm of the egg gives a very definite orientation to the various changes initially set up. When we view cell division from the standpoint of our knowledge of colloids we come to the conclusion that division is brought about by changes in the degree of swelling of the plasma-colloids, which are of course not equally distributed throughout the cytoplasm, but localized. Again, changes of surface tension in the fluid cytoplasm are certainly concerned in cleavage, for the surface tension is lowered by the raising of the water-content of the protoplasm at the poles of the division. This difference of surface tension is balanced by constriction of the cell: on the one hand, alteration of surface curvature is a consequence of altered surface tension; and, on the other hand, unequal radii of curvature (such as are shewn by the different regions of the surface of a dividing egg) must lead to an equalization of tension differences; because, as is well known, surface tension is inversely proportional to the radius of curvature (see Fig. 5). For the complete separation of the two halves of the cell forces of contraction may further be necessary, which are produced by local gelation in its interior (*vide* spindle and asters in the cytoplasm).

Now differences in the colloid phase-system of the cytoplasm and in the surface tension can be caused by local condensations and separation of colloids. In this way the action of chemical reagents, of hypo- and of hypertonic solutions, and of raised temperature is explained; for all these, according to the specific condition of the egg, lead, either to local condensations or gelations, or to swellings; so that differences of surface tension arise, and a cause is suggested of the formation of the cleavage spindle, in which local gelations are certainly concerned. Whether it be separation or gelation that is first set in motion, the final result is the same. Owing to the unstable nature of the cytoplasm, mechanical vibrations are evidently sufficient to produce separation, and such vibration may well be of importance in the case of the pricking experiments.

A comparison of these conditions with what happens in natural insemination is not at this point possible; but here other observations will help us further. It is well known that by the action of cell-sap and tissue-fragments cell-division may be provoked. Thus it is possible in plants, by pinching the ovary, to cause the formation of parthenogenetic adventitious embryos. Apparently from the broken-down cells substances arise which stimulate the egg to divide. These substances have been called "wound hormones," but they are better called *necrotins*, since they are substances produced by necrotic tissues. Now it is quite possible that when the egg is pricked with a fine needle necrotins of this kind are formed, and the more possible since a relatively large extrovate is produced, which completely degenerates. Further, the effect of the pricking is enhanced if cytoplasmic substances are inoculated into the unfertilized egg. For this purpose one may, with a needle, inoculate blood, lymph, minute tissue fragments and the like into the egg; or, before pricking the egg, it may be painted with a suspension of pulped testis in chloroform water, or with defibrinated frog's blood. A much higher percentage of eggs undergoing cleavage and gastrulation is so obtained than by pricking without this inoculation of formed substances. It is cellular elements especially which, when introduced into the egg, favour development. Since blood corpuscles used in this way have the same action as dead sperms, there can be no question here of a specific action of the male germ-cell.

The study of colloids has shewn that one colloid may cause the gelation of another, that is to say, may produce a change in its condition of swelling in the direction of condensation. The cytoplasm is—from the physico-chemical point of view— a colloid, and we may assume that the formation of the asters and of the spindle in cell-division is the expression of a localized gelation. From this point of view the initiation of cleavage by the inoculation of cytoplasmic material appears to result from the action of one colloid upon another: to be, in fact, a purely physico-chemical process.

As regards normal insemination, and the activation of the egg by enucleate spermatozoa, we may draw the following conclusions: the penetration of the sperm can be compared to the prick of a very fine needle; even if the injury to the protoplasmic surface is very slight, the possibility of the consequent

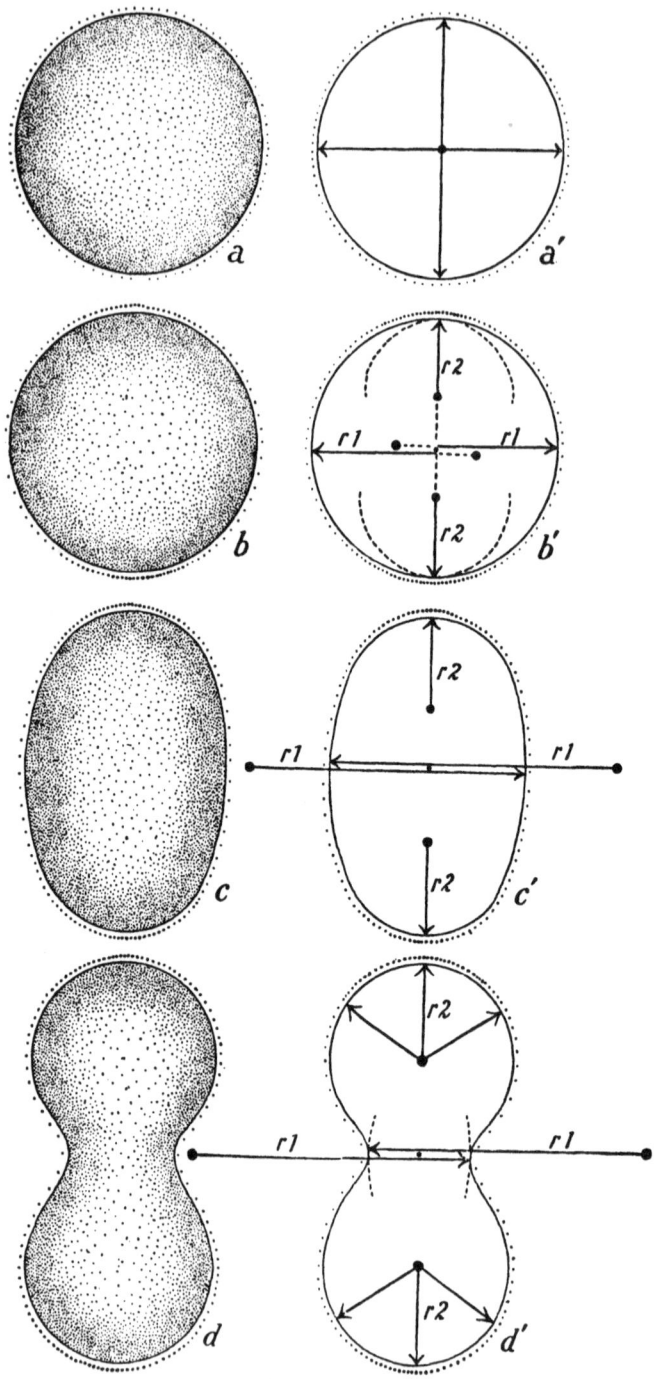

FIG. 5.
[For inscription, see foot of page 57.]

formation of necrotins is not to be dismissed. By the contact of the cytoplasm of the sperm with the surface of the egg, local changes may be caused in the surface-tension of the latter. Again, inoculation of foreign protoplasm into the protoplasm of the egg is comparable with the penetration of the substances of the spermatozoon. Local gelations, separations, and changes in surface tension can appear not only as results of insemination but of artificial activation also; the specific structure of the egg directs these local processes into topographically and functionally correct channels, so that division of the cytoplasm, and consequently of the nucleus, is set in motion. It remains to be seen in what ways this conception of activation will be modified by further research; provisionally these suggestions give at any rate a complete picture.

Looked at in this way cleavage appears as a mechanical process taking place in a plasma-colloid. It must be pointed out, however, that this gives by no means a complete physico-chemical explanation of cleavage in its relation to development, but accounts only for the physical mass-movements which produce the division of the cell. With these movements and displacements which concern the process of constriction and thus belong really to kinematics, there are involved certain biological considerations which should not be overlooked when envisaging the matter from the side of colloid physics. It will appear, moreover, that insemination not only leads to activation of the egg but is instrumental in causing fertilization, and that the nature of this must be judged from special standpoints.

Fig. 5.—Diagrams to explain the action of local differences of surface-tension.

a and *a'*, surface-tension the same all over (shown by the outer, evenly-spaced dots); surface with only one radius of curvature; hence spherical form.

b and *b'*, difference in surface-tension—increased equatorially, reduced at the poles of the sphere (shown by unequal distribution of outer dots); results: tendency of the radius of curvature r_1 to increase, of r_2 to diminish.

c and *c'*, compensation of tension-differences by change in radii of curvature (r_1 and r_2); form no longer spherical but elongated, the curvature of the polar areas being greater, that of the equatorial less.

d and *d'*, increase of differences in tension and curvature; the system is dumb-bell shaped, i.e. constriction has begun.

III. The Relations between the Two Gametes and the Nature of Fertilization

In assessing the real nature of fertilization the first matter to be decided is the relation of the two gametes to one another. We know already from observation that each of the two gametes contributes the same chromatin elements to the formation of the zygote nucleus. That being so the nuclei of the ripe egg and of the ripe sperm would appear to be equivalent. The cytoplasmic contributions of the two gametes to the fertilized egg are, from the morphological point of view, certainly different; whether they are also essentially different cannot here be decided. Genetic research has proved that, with the exception of very special cases, the egg and the spermatozoon are equally capable of transmitting the characteristics of the race, and that in the fertilized egg all the rudiments are present in duplicate, from the mother on the one side, from the father on the other. The two germ-cells must, according to this, be equivalent.

Analytical embryology is able to demonstrate the relative value of the germ-cells by testing their developmental capacity. Every egg is able of itself to produce a complete normal individual without the intervention of a spermatozoon. The question is narrowed down, then, to that of whether the same can be said of the male germ-cell. We can say at once that that is not the case, and for a reason that holds good for at least most kinds of spermatozoa. In order that embryonic development may take place cell divisions are necessary; these definitely demand a certain volume of cytoplasm, to make possible the division of the nucleus and the functioning of the other mechanisms of cell division. Quite obviously, however, spermatozoa lack this amount of cytoplasm. It may be asked whether there do not perhaps exist sperms which would be able to divide, and therefore to perform something analogous to the cleavage of the egg; but no such case is at present known, nor does its existence appear very probable.

That being so, the suggested comparison is narrowed down to

a comparison of the nuclei alone of the two gametes. Both in normal and in artificial parthenogenesis the nuclear material of the offspring is derived from the egg only. The problem is, then, to produce individuals whose nuclei are derived entirely from the sperm-nucleus, without any contribution of nuclear material from the maternal side.

This problem may be successfully attacked in various ways: first by producing *merogony*, that is to say the development of a fragment of an egg. Sea-urchin eggs, when shaken for some time in a test-tube, readily break into fragments, of which only one fragment from each egg may happen to contain the nucleus, the others being enucleate. Such fragments can now be inseminated and are capable of development. Though their nuclei are derived solely from spermatozoa they give rise to cleavage-stages and even larvæ. These last differ from normal plutei only in being smaller. The diameter of the egg-fragment with which the sperm unites may be very small. If it be even $\frac{1}{20}$ of that of the egg a normal larva can arise, though, naturally it will be very small. The fact that development goes no farther than the pluteus stage is chiefly due to the fact that it is extremely difficult, even after normal fertilization to rear sea-urchin larvæ in the laboratory past their metamorphosis.

In the way we have just described it is not only possible to obtain larvæ from enucleate egg fragments by insemination with sperm of the same species, but one can also, within limits, produce a kind of cross between different species. This crossing, however, differs from a true crossing or hybridization, in which egg and sperm nucleus unite, in that one species supplies only the cytoplasm, the other only nuclear material. It is not possible to make any such combination at will; only in certain cases are they successful; and the less nearly related the species employed the less perfect is the development of the combination. In certain species (e.g. egg-fragments of *Parechinus microtuberculatus* with sperm of *Paracentrotus lividus*) even larvæ can be obtained; in others, on the other hand, cleavage starts but development very soon comes to a standstill. The male pronucleus is therefore itself able to perform the developmental functions of a zygote nucleus—not with any kind of cytoplasm, but only with cytoplasm which is to some extent in accord with its own specific character.

The larvæ obtained by merogony can be recognized as such by the size of their nuclei, for these nuclei are derived from only one gamete nucleus, and are therefore smaller than in normal larvæ. Larvæ produced without the assistance of an egg-nucleus can, in fact, always be recognized by the smaller size of their nuclei.

Another method of producing individuals with only paternal nuclear material consists in fragmenting the egg not before, but after insemination. Only those kinds of eggs in which polyspermy is normal can thus be used. The entrance of several—in some cases of many—spermatozoa is the rule in the case of very yolky eggs, such as those of Amphibia, Birds, many Fishes, Insects, etc. In all these only a single sperm is actually concerned in fertilization, since a single male pronucleus fuses with that of the egg to form the zygote nucleus. The egg of *Triton* is normally polyspermed. Since after fertilization the egg-nucleus lies eccentric, it is fairly easy with a loop of hair to constrict the egg so that the zygote nucleus lies in one-half of it and one or more sperm-nuclei in the other. The egg is thus divided into a diploid half and a haploid, merogonous half. Both halves develop, though the latter is slower than the former. From the half with the normal, diploid nucleus a typical larva arises. The haploid half also produces a larva which, though obviously dwarfed, appears otherwise normal, and may pass through its metamorphosis. The nuclear size is, in this case also, markedly different; the smaller haploid, nuclei betray their origin from a single pronucleus—that of the sperm.

Finally, an individual can obviously be produced without the help of the egg-nucleus by enucleating the egg and then inseminating it normally. The removal of the egg-nucleus can be performed in various ways. In Holothurian eggs it has been possible by the use of hypertonic solutions of salts to produce amœboid movements which result in the extrusion of the nucleus; the enucleate egg can then be inseminated and reared as far as cleavage. But the most successful procedure for enucleation is the one mentioned above, in connexion with the enucleation of spermatozoa—that is to say their irradiation with radium. The final result is the same whether the nucleus be actually taken out of the egg, or whether it be damaged so much as to make it no longer capable of being fertilized or of taking part in development. When the eggs of *Triton* are irradiated at a distance of 3 mm. with mesothorium (= 51 mg. of pure radium bromide) for more than five minutes one can be certain that the egg-nucleus is killed. Insemination of these eggs with normal sperm yields, in a sufficiently large number of cases, larvæ which will live for weeks. The larvæ are dwarfed and show a lessened viability, exactly like the parthenogenetic larvæ in the reciprocal experiment; but for us the really important fact is that development has been achieved without the participation of the egg-nucleus. Measurement of the nuclei and counts of the chromosomes demonstrate conclusively that the sperm nucleus alone, and its derivatives, formed the basis for the cell nuclei of these larvæ. Just as in the case of artificial parthenogenesis, phenomena of regulation occur that modify the chromosome-number, and occasionally larvæ are observed in which the normal diploid number is found; irregular chromosome numbers may also appear as a result of imperfect regulation.

It follows from this that either the male pronucleus or the female pronucleus will separately suffice for the process of development. Though it certainly cannot be said that an individual can arise equally from the egg or from the spermatozoon, yet it is true that, as far as the behaviour of their nuclei is concerned, the two kinds of cells appear to be almost exactly equivalent. On the basis of the above experiment this equivalence cannot be postulated for the cytoplasmic parts of the two kinds of cells. We cannot on that account, however, simply neglect the sperm cytoplasm. If there were present in the sperm a larger amount of cytoplasm, which mixed with that of the egg, this would certainly not be without importance in the ensuing processes of development. The fact that the amount of sperm-cytoplasm is usually very small is not sufficient to justify its general neglect, since great importance is ascribed to quite minute portions of the nucleus. Nothing definite can be asserted about the essential role of the sperm-cytoplasm; the assessment of its relative importance in comparison with the egg-cytoplasm must be left to the future.

Apart, however, from this question we must recognize the fact—which, indeed, follows from the phenomena of heredity—that male and female reproductive cells are equivalent to one another. This gives us a criterion for the diagnosis of the process of fertilization. What is essential in this process is not the activation of the egg, but rather the union of two separate systems, very similar as regards their potency, to form a new individual. This union which, in general, appears to be possible only between single cells must *ipso facto* coincide with the beginning of embryonic development—with the activation of the egg. Here we are dealing, then, not with a simply external and mosaic-like apposition of two component systems to form an outwardly homogeneous complex, but with the origin from two individual integral systems of a new unit and whole, even though the dual origin of the new organism so produced is still recognizable. It will be shewn later that such a fusion of two individuals to form a single organism can be brought about experimentally in other ways. The feasibility of such a fusion, as indeed of the union at fertilization, reveals the presence of a specific organismic faculty which, in the last analysis, cannot be referred to any organization either mechanistically or morphologically explicable. It is deeply rooted in the special idiosyncrasy of the living system, and gives us some indication of the direction in which we must seek for the essential nature of life.

CHAPTER III

PRESUMPTIVE ORGAN-REGIONS AND THE KINEMATICS OF EARLY DEVELOPMENT

I. FORMATIVE MOVEMENTS AS A FUNCTION OF THE WHOLE GERM

By the process of cleavage the germ is transformed from the unicellular to the multicellular state. The embryonic cells which thus arise may appear to be almost identical or may shew from the outset obvious and sometimes quite considerable differences. These differences, whether they declare themselves during cleavage or not until later on, stand in a natural relation to the developmental fate of cells and groups of cells—or, in more general terms, to the fate of particular germ-regions. The fact that they may appear either during or after cleavage leads along two lines of deduction to important conclusions. These, though here considered in general outline, will only be properly understood in detail later, after the discussion of other results. First, then, cleavage—as such—cannot be the cause of the differences between individual cells or germ-regions; neither, in the second place, can the cleavage of the egg be regarded as always being the first step in its development, though in many cases it may give that impression.

The fundamental processes of development must be taken to be essentially the same for all groups of animals; it is the outward form only that can vary according to the initial constitution of the germ. If therefore germs exist which become multicellular without the appearance of marked differences between the individual cells, cell-division itself cannot be the cause of the differences between the germinal regions; at most it can render the differences sharper. These differences must have another cause, and one which in many cases was at work before the beginning of cleavage; in other words actual processes of development occur before the division of the egg. The nature of these processes will not concern us until later. It is sufficient to say here that they involve the appearance of local differences in the protoplasmic mass of the fertilized egg—differences which must be looked upon as the expression of a more or less complete determination of developmental fate in the different regions. In any case cell-division does not bring about determination: it is only a means by which development is carried out.

In the majority of cases the individual cells of the germ, and their descendants, do not lie to begin with in their ultimate functional position, but undergo considerable displacements, which bring them into the positions where their destiny is fulfilled. But that is not to say that individual cells migrate of their own accord; on the contrary it is mass-movements of the whole germ that are concerned, even though in certain cases the movement of single cells is what we observe. This is notably true when such displacements occur on a considerable scale in quite early stages—i.e. when few cells are present, or in germs which in any case are composed of a strictly limited number of cells.

We may take the processes in the Nematode egg as an example. The egg of *Ascaris* (p. 73, Figs. 18 to 23) produces, by its first cleavage, two slightly unequal blastomeres, This inequality is expressed by the fact that their directions of cleavage are different. In this way there is formed the characteristic T-stage of the Nematode egg (Fig. 19), in which two cells form the cross-piece and two cells the tail of a T. This form of the germ does not long persist; by a swinging round of the tail towards the hinder end of the germ it changes into the rhomboid form (Fig. 20), which itself, after further cell-divisions and displacements, disappears (Figs. 21 to 23). As can be seen by a comparison of the figures, minor displacements and changes of form of the remaining cells are connected with this swinging round of the tail. It is particularly obvious in this case that the single cell is responsible for the movement, because the whole mass of the germ is limited at this stage to a few (4) cells. Nevertheless here, and in other examples of change of form in early development, we are dealing with the activity of the germ as a whole. In the case of *Ascaris*, this is easily seen in later stages, during gastrulation. It is demonstrated much more clearly, however, in germs which carry out such movements only after they are composed of many cells; observation of the mass-movements is then no longer embarrassed by the single cells, the importance of which is diminished.

In such cases simple observation of the living germ, or a survey of separate stages of the young embryo chronologically arranged, no longer suffices. It is comparatively easy to follow the movements and displacements of a germ consisting of few cells, especially when the individual blastomeres are, from the beginning, different in appearance, for the germ is easily examined and has clearly marked differentiations. In the case of a many-celled germ, however, where the cells are more or less alike, it is impossible to follow exactly the kinematics of early development without the help of experiment. The method of doing this which has been most successful consists in local *intra-vitam* staining, which at the same time forms the most useful means of studying the presumptive organ-regions.

Living cytoplasm has the peculiarity of taking up and retaining certain stains, such as Nile-blue sulphate and neutral red. This makes it possible to mark the surface of the unsegmented egg, or of the young germ, in any way desired, and allows the course of the movements to be traced with great ease.

The use of this method has been carried farthest in the case of the egg of *Triton*. The movements naturally appear most strongly marked during gastrulation. When, for example, a ring of stained patches is made on the equator of the morula of *Triton* (Figs. 6-10), distinct changes of shape in these marks appear during gastrulation—they are strongly elongated, and give the immediate impression that cell material is streaming in over the blastopore lip. Another change of form consists in the fan-like spreading out of the ventral marks during the formation of the medullary folds (Fig. 9), which again points to displacements of material. Finally one can follow the position

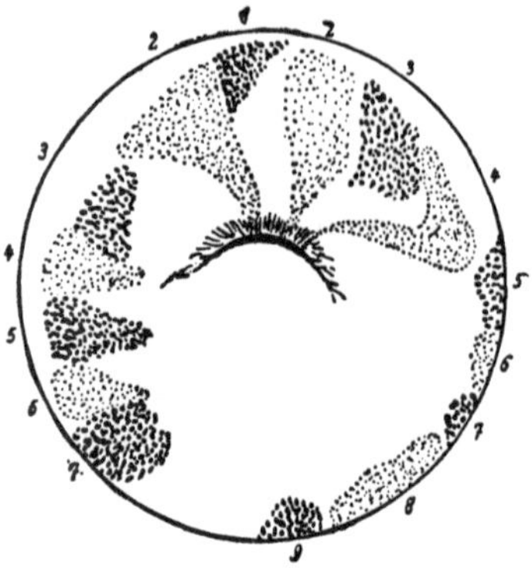

FIG. 6

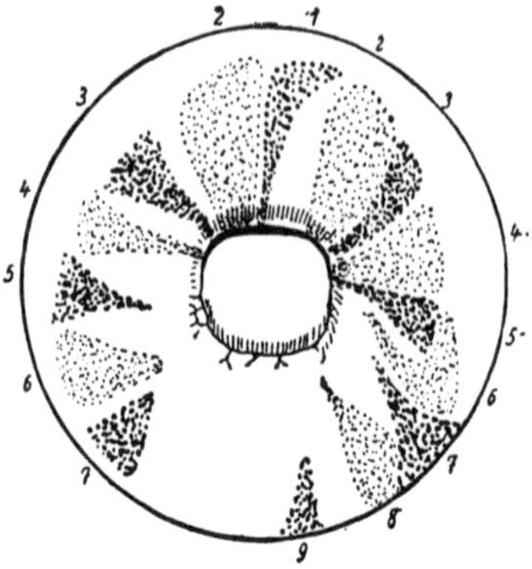

FIG. 7

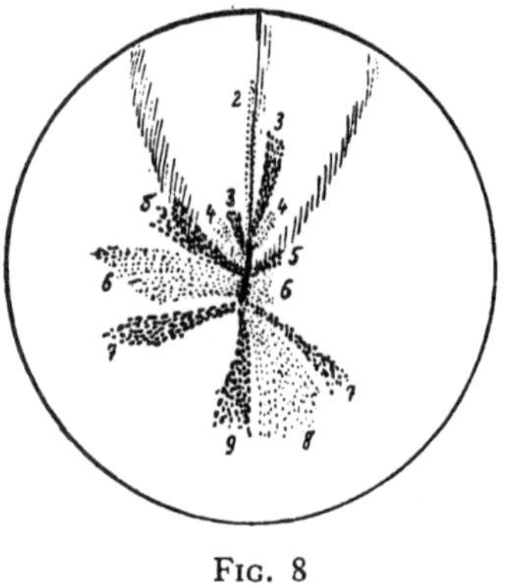

FIG. 8

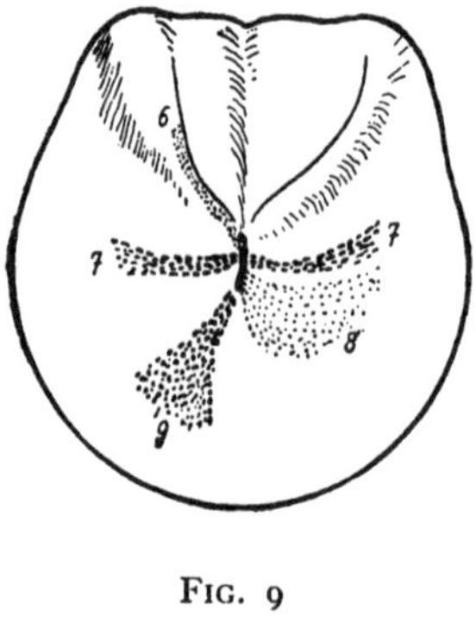

FIG. 9

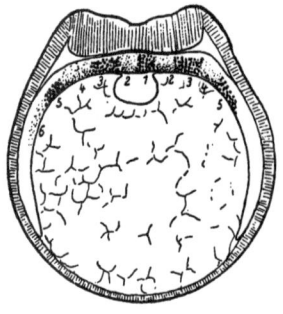

FIG. 10

FIGS. 6 to 10.—Behaviour of stained patches made on the equatorial region of the morula of *Triton*. Inward streaming of germinal material during gastrulation accompanied by elongation of the marks, and the positions in which the stained portions finally lie. (Coarse stippling—Nile-blue sulphate; fine stippling—neutral red.) Fig. 6, beginning of invagination of the archenteron, and movement of the marked marginal zone towards the blasto-pore. Fig. 7, in-flow of material. Fig. 8, convergence of the marked germinal material towards the dorsal region. Fig. 9, fan-like arrangement of the marks on the ventral ectoderm during the appearance of the medullary folds. Fig. 10, position in which the marked dorsal parts come to lie in the chorda-mesoderm plate. (After W. Vogt.)

of the marked germinal regions far into the development of the embryo.

In all these movements we have to do not with the wandering of individual cells but with material which, approaching the edge of the blastopore and flowing in around it, moves as a whole with a single impulse; one might well describe what happens as an amœboid movement of the whole germ. The single cells are passively carried along. With exquisite clarity we see here that what supports the entire process of development is not the individual cell but the germ as a whole. It appears from the results of such marking experiments that the rate of division is very important in the formative processes of early development, while the direction of division is hardly important at all.

The significance of this conclusion is not in the least prejudiced by the fact that in certain circumstances, as we have seen, the movements of individual cells, resulting from special conditions present in the system, force themselves upon our notice. Thus, in spite of a subdivision into many cells, unity of the germ is preserved, as is proved by the fact that its formative movements function as a whole.

II. Presumptive Organ-Regions

1. General Characters of a Presumptive Organ-Region

In the normal course of embryonic development any given germ-region gives rise to a perfectly definite organ or group of organs. Such a germ-region, therefore, can be called the presumptive region of this organ or group of organs—in short, a presumptive organ-region. Speaking purely descriptively, each of these regions is derived from a definite protoplasmic region of the unsegmented egg. That is not to say, however, that the fertilized egg is made up of a mosaic of organ-rudiments, each organ being directly preformed in a special portion of the protoplasm; development does not generally depend, as will appear more clearly later, on the direct preformation of differentiations. The establishing of these regions has at first no more than a purely descriptive and topographical meaning.

In many cases we cannot at first recognize different regions in the protoplasm of the egg; at most there exists at an early stage a polarity in the direction of the primary axis of the egg, passing from the animal to the vegetative pole, but which often is not manifested morphologically. In every other way the cytoplasm of the egg appears homogeneous throughout, or isotropic; the cleavage cells also are generally markedly similar. If the morphogenetic movements of early development are followed, however, it is seen that a locally defined protoplasmic region is converted into a definite germ-layer, or into a definite organ-rudiment, so that such a part of the egg can be called the presumptive region of this organ-rudiment, or of the organ in question. It need hardly be said that the position in space of the presumptive region may be other than that of the organ which arises from it. That the presumptive organ-regions are not simply the organs themselves in a primitive form is proved by the fact that modification of the development of such a region leads to the production of an organ totally different from that which normally arises.

In other cases, plainly circumscribed local differentiations of the cytoplasm are already recognizable in the unsegmented egg, and these regions are transformed severally into definite organs and organ-regions. Between these two extremes stand those kinds of eggs which before cleavage appear almost completely isotropic, but in which during cleavage differences arise among the blastomeres. It is therefore not very difficult to ascertain the fate of particular regions of the egg, or of particular blastomeres, since these differences in cytoplasmic character of the regions or blastomeres make it possible to determine the positions in which they come to lie. It is in this way that the presumptive organ-regions of the egg and of the young germ have been discovered.

To ascertain the positions and limits of these presumptive regions is of interest chiefly because the fixing of their topography is an important preliminary to the real analytical study of development. Naturally this investigation has not been carried out in all cases with the same precision and completeness. We shall content outselves with three examples.

2. Presumptive Organ-Regions in Mosaic Eggs

The eggs of Ascidians are among those which, even before cleavage, show a characteristic division of the cytoplasm into definite areas. The unfertilized egg of *Cynthia partita* shows on its surface no topographical differences, its superficial layer of cytoplasm having yellowish pigment-granules evenly distributed throughout it. But after the penetration of the spermatozoon its homogeneous character changes. The yellow cytoplasm contracts into a definite area and forms now the so-called *yellow crescent* (Fig. 11, *g*). As the further history of this area shews, we have here the presumptive mesoderm region (*m*), which later becomes divided into the region of the mesenchyme and that of the myoblasts. At the same time a region of clear cytoplasm appears in immediate apposition to the yellow crescent (Fig. 11, *h*). This, even before the completion of the first cleavage, spreads out more widely over the animal half of the germ to form the region of the presumptive epidermis (*e*). Thus in the Ascidian egg the presumptive organ-regions are not laid down once and for all from the beginning, but their separation and localization come step by step, even though it be at a very early stage. The formation of the two-celled stage continues this: we still find in the animal half of the germ the region of presumptive epidermis (*e*), but in the vegetative half three regions are plainly distinguishable. One of these, the mesoderm region (*m*), was already seen in the unsegmented egg; opposite to it lies a clearer region of similar extent and shape, the chorda-neural region which later gives rise to the notochord and nervous system—division of the two presumptive rudiments has not yet been effected at this stage. The third region takes over the vegetative half of the germ, and in particular covers the vegetative polar cap; it consists of thickly

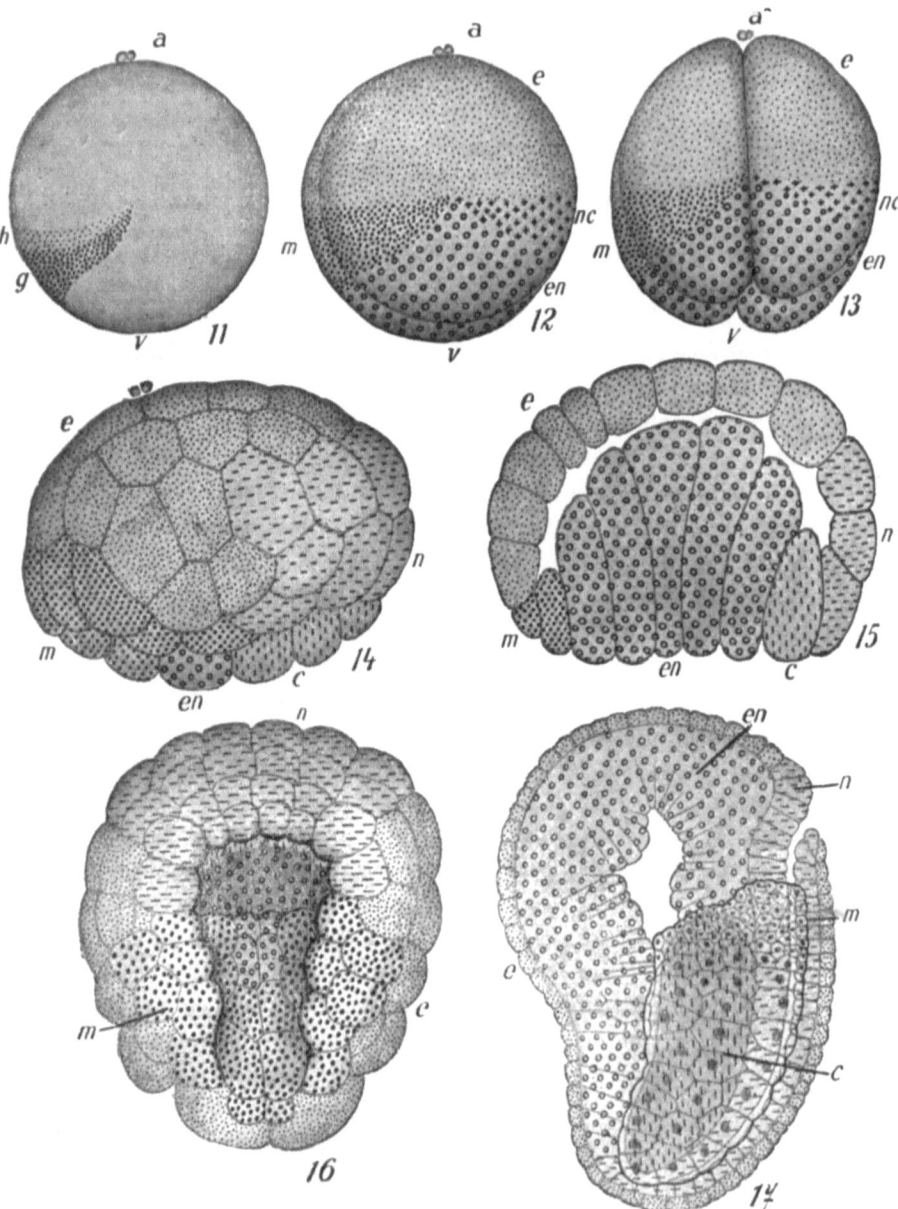

FIGS. 11 TO 17.

[For inscription, see foot of page 71.]

granular material, rich in yolk, and constitutes the region of presumptive endoderm (*en*).

Sharp boundaries do not appear between the individual protoplasmic regions, and their separation and arrangement are not always carried out in the manner described, but may be spread over further cleavage stages. Once they are established, however, further cleavages have no effect on the general arrangement of the different cytoplasmic regions (cf. Fig. 13, four-cell stage). Even at the beginning of gastrulation (Figs. 14, 15) the relative positions of the presumptive organ-regions are still exactly maintained, except that the neural region (*n*) is now separated from the chordal region. The progress of gastrulation (Fig. 16) naturally involves displacements of the regions corresponding to organs and germ-layers, and one can now distinguish, within the limits of the mesodermal region, presumptive muscle from what later becomes the mesenchyme. Fig. 17 shows an embryo in which the individual organ-rudiments are already well developed.

To sum up, we can say that in the Ascidian egg, at least as early as the two-celled stage, the regions from which originate the epidermis, the mesoderm, and the endoderm, the nervous system, and the notochord can be distinguished from one another by differences in the character of the protoplasm, and their topography can already be defined. And though the different regions are formed only during the course of cleavage, yet this last has itself no effect on the division and topography

---

FIGS. 11 to 17.—The presumptive organ-regions in various stages of development of *Cynthia partita* (Ascidian); semi-diagrammatic (from Conklin's figures of the cleavage). Fig. 11, egg before cleavage. Fig. 12, 2-cell stage. Fig. 13, 4-cell stage. Fig. 14, beginning of gastrulation. Fig. 15, sagittal section of the young gastrula. Fig. 16, further stage in gastrulation. Fig. 17, older embryo shewing first rudiments of organs. Figs. 11 to 15 and Fig. 17 shew the germ from the left side, Fig. 16 from the lower (i.e. blastoporal) surface.

a animal pole; *v* vegetative pole; *g* yellow crescent; *h* region of clear cytoplasm; *e* presumptive epidermis (ectoderm); *m* presumptive mesoderm-region; *en* presumptive endoderm-region; *nc* presumptive chorda-neural region; *n* neural region (neural tube in 16); *c* notochordal region (notochord in 16). Presumptive epidermis, finely dotted; mesoderm, coarsely dotted; endoderm, small circles; chorda-neural region, crosses; neural region, horizontal lines; notochordal region, vertical lines.

of the protoplasmic regions. We shall now see the same thing proved in a very different case.

In the egg of *Ascaris* the presumptive organ-regions are not defined before cleavage, and only after several divisions does this to a certain extent take place. The polarity of the egg is already present during its growth period in the egg-tube; that side of the egg which is attached to the so-called rhachis becomes the vegetative pole; the free side, the animal pole. There are no obvious material differences between the two poles—at least at first; in the ripe egg the arrangement of substances shews a gradient. As we have already seen, the first cleavage divides the egg into two unequal blastomeres (cf. Figs. 18 and 25). One of these cells is the so-called primordial soma cell (S_1 or AB, according to the usual nomenclature). The other, the so-called first stem cell (P_1). The former cell is already a presumptive organ-region, and corresponds to a part of the ectoderm with its derivative organs, while the second cell possesses no such definite association. By the next division there arise from S_1 the cells A and B; from P_1 the second stem cell (P_2) and the primordial soma cell S_2 ($EMSt$). The germ has now the T form we have mentioned (Fig. 19), and this at once passes into the rhombus shape (Fig. 20). The descendants of S_2 give rise later to endoderm, mesoderm, and the material for the stomodæum, but it is impossible as yet to distinguish these three regions. The further course of division and the significance of cell-lineage in relation to the formation of organs is easily seen by referring to Figs. 21–24 and to the diagram, Fig. 25.

In the division of S_1, S_2, S_3, S_4, the so-called *diminution of chromatin* occurs which is characteristic of the somatic cells, while in cells P_1, P_2, P_3, P_4, it is absent. In this diminution portions of the ends of the chromosomes are cast off. P_4 is the primordial germ-cell, and receives the undiminished original chromosomes. The cell-lineage P_1 to P_4 is the so-called *germ-track*—that is, the sequence of cell-generations through which the primordial sex-cell, and therefore also the definitive germ-cells themselves, are connected with the egg. Special importance

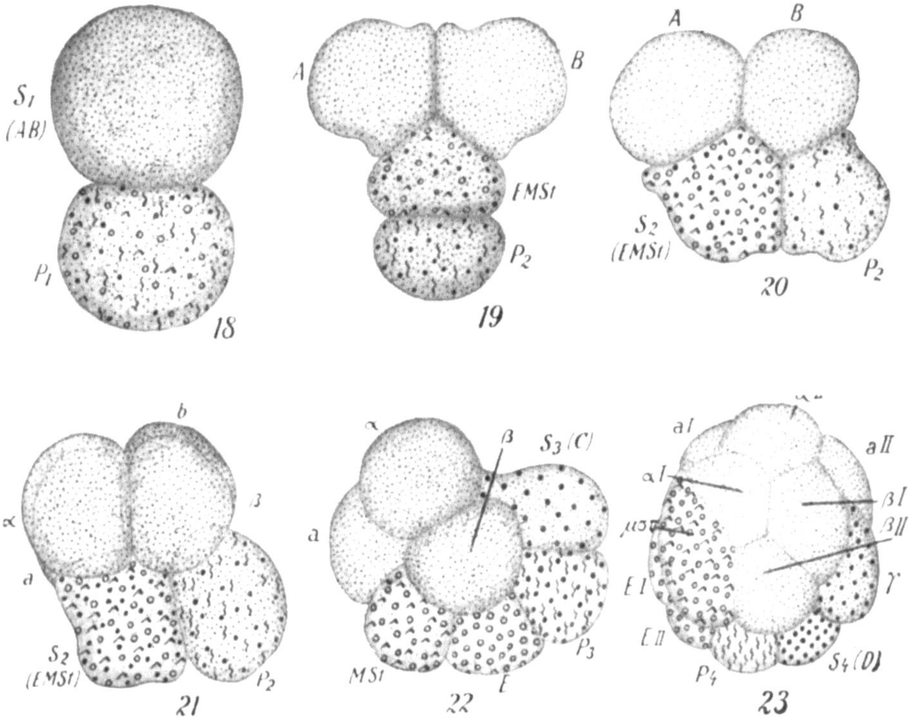

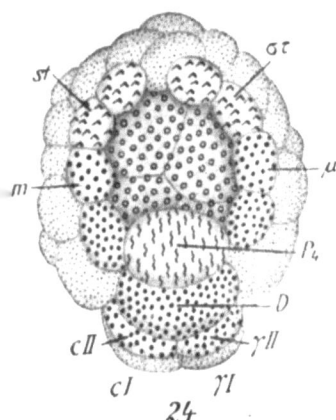

Figs. 18 to 24.—Cleavage stages of *Ascaris* to explain the topography of presumptive organ-regions in the young germ; semi-diagrammatic (based on the figures of Boveri and zur Strassen). Fig. 18, 2-cell stage. Figs. 19 and 20, 4-cell stage. Fig. 21, 6 cells. Fig. 22, 8 cells. Fig. 23, 16 cells. Fig. 24, beginning of gastrulation. Figs. 18 to 23 shew the germ from the left side; Fig. 24 from the ventral or blastoporal side. For the way in which the cells of later stages are derived from those of earlier stages, see the diagram, Fig. 25. Presumptive germ-layers and organs are shewn by marks similar to those used in Figs. 11 to 17: ectoderm, small dots; mesoderm, coarsely stippled; endoderm, clear circles; stomodæal cells, ⌃ ⌃; cells of germ-track, short sinuous lines. Cells whose descendants pass severally into more than one germ-layer shew the fate of these descendants by their mixed marking. The presumptive regions in such cells are still undefined, and the various marks are therefore scattered haphazard; but it must be noted that a separation is present before cleavage begins.

was formerly ascribed to this germ-track, but it is clear from what has been said above that all presumptive organ-regions, and hence all the cells of an organ, are derived through their

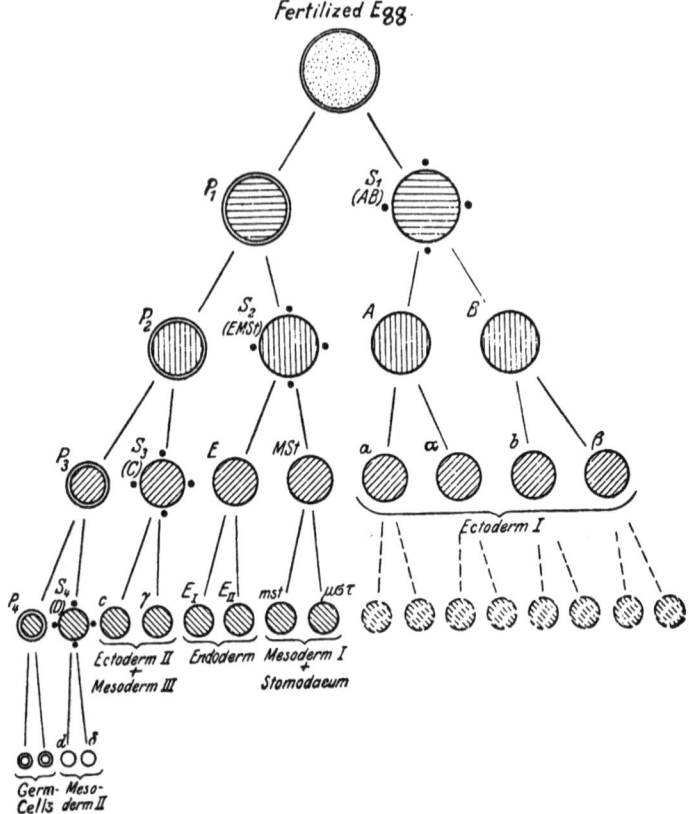

FIG. 25.—Diagram of the cleavage of the egg of *Ascaris*. Circles with double outline, cells with intact chromosomes. Single circles, cells with diminished chromatin. Single circle with 4 black dots, cells in which chromatin-diminution is occurring. All the cells of a particular cell-generation are shaded alike. 2-cell stage: P_1, S_1 ($= AB$). 4-cell stage: P_2, S_2 ($= EMSt$); A, B. 8-cell stage: P_3, S_3 ($= C$); E, MSt, a, α, b, β. 16-cell stage: P_4, S_4 ($= D$); c, γ, E_I, E_{II}; mst, $\mu\sigma\tau$, and the daughter-cells of a, α, b, and β (dotted circles).

proper cell-lineage from the egg, and that the germ-track is only one special case of this.

In the 6-cell stage (Fig. 21) no further discrimination of presumptive regions is as yet possible, but in the 8-cell stage

(Fig. 22) it is; for at this time the presumptive endoderm region is separated off (cell E). Several regions of the germ still, however, shew a mixed character in regard to their presumptive value. This is true also of the 16-cell stage (Fig. 23), but a more marked separation can now be seen in these regions. Part of the presumptive mesoderm is now present as a single homogeneous region (cell D) and so also is the parent cell of all the germ-cells (P_4). After further divisions the demarcation of the presumptive germ-layer regions (using this expression with a certain reservation) and of two presumptive organ-regions—those namely of the stomodæum and of the generative parts of the reproductive organs—becomes possible.

We have, then, in the early gastrula stage (Fig. 24) presumptive regions for the following rudiments: (1) ectoderm cells comprising the whole dorsal side of the germ; (2) endoderm cells, already beginning to sink in; (3) mesoderm cells lying at the hinder end of the germ and at the ventral edge of the ectoderm region; (4) stomodæum, whose cells lie anterior, at the ventral edge of the ectoderm region; (5) the parent cell of the germ-cells (P_4), lying at the hinder end of the germ, in front of the mesoderm cell. The derivation of definite organs from definite parts of the young germ, and the topography of these parts, is thus accounted for.

3. Presumptive Organ-Regions in Regulative Eggs

Apparently the topographic relations of the presumptive regions in the germ of *Ascaris* are only produced step by step. But it should be clearly understood that before a particular cleavage occurs the corresponding stage must in every case have been reached; this is not patent to observation alone because the cytoplasm of this germ has a uniformly homogeneous appearance (contrast here the germ of *Cynthia*). Within individual cells it is therefore impossible to distinguish one region from another, so that we have at first cells of an indifferent, mixed character; but it is conceivable that by experimental methods we might succeed in demonstrating this distinction, and that the topography in question could thus be specified more exactly than is at present possible, and in much younger stages.

In other germs where the cells appear very similar the different regions can be successfully discriminated by the local *intra-vitam* staining of special regions of the egg, as described above for the investigation of the formative movements of early development. In this way, by marking systematically with stained patches certain points on the surface of the egg or of the blastula, and tracing each mark to its position in the embryo, one is able, finally, to establish both the extent and the relative position of the presumptive organ-regions on the surface of the unsegmented egg, the blastula, and the gastrula. It is in the case of the germ of *Triton* that the topography of these regions has been most completely and exactly determined.

By this use of marks, made before and during cleavage, we find that the topography of these regions is defined unusually early in the morula, the blastula, and the young gastrula. The subdivision of the egg during cleavage into cells—and finally into many cells—does not substantially alter the arrangement of these protoplasmic regions; and this we have already seen to be true of the eggs of Ascidians. We have, then, a further confirmation of the fact that cleavage is not responsible for the production of the presumptive organ-rudiments.

Now, in the egg of *Triton*, the topography we are speaking of has not been so exactly determined in earlier stages of the germ as in the young gastrula (Pl. I, Figs. 26–29). Five regions can be distinguished with certainty on the surface of the blastula. The first of these is that of the ectoderm or the epidermis (Fig. 26, *e*). It reaches not quite to the animal pole, spreads somewhat far into the vegetative half of the germ, and is bounded elsewhere by the region of the medullary (neural) plate and that of the presumptive mesoderm. The presumptive medullary material forms a wide band, which runs out to a point on either side of the germ, stretching far round the sides of the blastula (Fig. 26, *n*). It includes the animal pole of the egg. Lying on the side which is towards the future blastopore we find the presumptive mesodermal and notochordal materials; the former encircles the whole of the part of the blastula between neural, ectodermal, and endodermal regions; the latter occupies a somewhat broad area of the median region of the germ (Figs. 26, *m* and *c*). The vegetative pole lies within the presumptive endoderm, which in the region where the anterior lip of the blastopore will be formed reaches higher than on the other side (Fig. 26, *en*). Boundaries which are diagrammatically precise cannot really be established between the different regions; their limits are always somewhat fluctuating.

In the earliest gastrula stage the presumptive areas can be determined more exactly (Figs. 27 and 28). In principle their arrangement is now exactly as in the blastula, except that the vegetative pole, as a result of incipient invagination, is already somewhat shifted towards the upper lip of the blastopore. The animal half of the germ is occupied principally by the materials for epidermis and medullary tube (Fig. 27, *e*, *n*); both regions pass downwards as far as the invaginating edge round which mesoderm and endoderm materials pass inwards during gastrulation. The animal pole is now found just within the medullary material. The presumptive mesoderm reaches right round the germ to the side opposite the upper blastopore lip; above this lip it is interrupted by the wide extension of the notochordal material, the wings of which spread far round the sides of the germ (Fig. 28, *c*). In the mesoderm region itself we can already distinguish between material for the somites (*m*), for the lateral plates (*m'*), and for the tail-bud (*m''*). The almost circular field of the gastrula which remains below is occupied by presumptive endoderm material (*en*). In the neighbourhood of the first in-tucking of the blastopore the distinction between notochord, endoderm, and mesoderm is still somewhat uncertain. It is here that the front end of the notochord lies, while its hinder end is formed from the region adjoining the medullary material. That a narrow, elongated organ, the notochord, is later formed from the widespread chorda region is explained by the fact that during its in-streaming round the dorsal blastopore lip there is a continual concentration of material towards the middle line; it thus comes to lie in a median position under the medullary material. This last also has its widely-spread arrangement transformed into an elongated one by convergent streaming.

Now, because we can determine, before gastrulation, the presumptive materials for a whole series of organs, it should not be supposed that these topographically determined regions are, so to speak, the actual organs in a primitive embryonic condition. That this is not the case emerges most clearly from the fact that their fate in development is by no means finally settled: it is, on the contrary, undecided and capable of being directed along lines very different from those which it would normally follow. This matter will be dealt with later.

Quite apart from the different outward appearances of cleavage—due to the varying direction of the cleavage planes, to the complete or partial cleavage of the egg, and so on—two special types of the process can be clearly distinguished, namely the determinate and indeterminate. Speaking in a purely descriptive sense we must count among the former those eggs and germs which at an early stage—before cleavage or at least early in cleavage—already shew the presence of presumptive organ-regions by differences in their protoplasmic regions or blastomeres. Among the latter must be reckoned those which, at least in quite early stages, shew no outward signs of the presence of such regions. The three examples given prove, however, that this difference cannot be fundamental, but is one only of degree.

From the point of view of analytical embryology too the types of cleavage are different. The determinate type is distinguished by the fact that the fate of individual blastomeres or germinal regions is already settled at a very early stage, while in the indeterminate type this specification only appears during later development. The purely descriptive difference between presumptive organ-regions is simply the outward expression of this chronological difference in determination. We shall see later that even this experimentally ascertained difference is not a primary one but only marks the two extremes of a continuous range of behaviour. Thus, descriptively, germs which shew strikingly obvious organ-regions at an early stage can be connected through intermediate forms so as to lead without discontinuity to those in which such obvious marks are lacking. As in Ascidians and Nematodes, a determinative type of cleavage is shewn by the majority of Molluscs, Annelids, and Ctenophores. Insects which were formerly reckoned as belonging to this type occupy an intermediate position. A type of cleavage which is relatively if not absolutely indeterminate is found in Amphibians, Teleosteans, Nemerteans, Amphioxus, and also in Mammals. These differences will concern us later when we come to deal with the problems of potency and determination.

III. Egg Axis and Cleavage Plane in Relation to the Developed Organism

The primary axis of the egg which passes from the animal to the vegetative pole, is marked in many eggs by a special disposition of substances; other forms lack this morphological relation, and the animal pole is then generally recognizable as the place at which the polar bodies are formed. If it be asked whether this axis, present in the unicellular stage, becomes any particular axis of the developed organism, it must be admitted in the light of our present knowledge that such is not generally the case. In the Ascidian egg the primary axis makes an acute angle with the long axis of the embryo, so that the original vegetative pole lies postero-dorsal and the animal pole antero-ventral. In Amphioxus the relations are similar, while in Amphibians (e.g. *Triton*) they are quite different. In this last case, as we have just learnt, the original animal pole lies within the presumptive medullary material, and thus corresponds to the floor of the fore-brain. Though the long axis of the embryo in this way coincides approximately with the primary axis of the egg, there is nevertheless a definite divergence; the egg-axis cuts the embryonic axis at an acute angle, so that the former deviates from the latter dorsalwards in front, ventralwards behind. In Nematodes there are considerable changes of position in the parts of the germ, so that a strict relation between the egg-axis and any characteristic axis of the animal cannot be established. This may well be the general case. On the other hand there are also cases where the position and orientation of the primary axis of the egg become those of a principal axis of the body. This is especially true of the Ctenophores, where the axis passing from the apical pole to the mouth coincides in position with the primary egg-axis, or, to speak perhaps more correctly, with the axis of the first cleavage plane. Something similar is found in Annelids. Here the cleavage-axis from the animal to the vegetative pole corresponds to the principal axis of the trochophore larva—taking that to be the axis from the apical plate to the anus. It is seen later that this almost—but not quite—coincides with the long axis of the adult worm. If we survey the whole range of developmental forms we must conclude that in the majority of cases neither the primary nor any other axis of the body is directly preformed as the primary axis of the egg, though sometimes it can be shewn that these axes do coincide. This fact alone goes some way towards proving that morphogenesis is only indirectly conditioned.

Division of the egg often presents a very regular picture in the arrangement of cleavage planes and hence of the first-formed blastomeres. In descriptive embryology especially—but also in experimental embryology—the question arises as to whether the first cleavage plane produces a definite plane of symmetry of the body. There are in fact cases where it does. The Ctenophores must first be mentioned in this connexion; in these the first cleavage plane is the same as the so-called stomach plane of the adult animal; the second cleavage, which is at right angles to this, corresponds to the tentacle-plane. This relation of cleavage planes to the principal symmetry-planes of the body is not, however, always recognizable, though occasionally such coincidence occurs. Fairly frequently in the Anura, for example, there is coincidence of the first cleavage

PLATE I

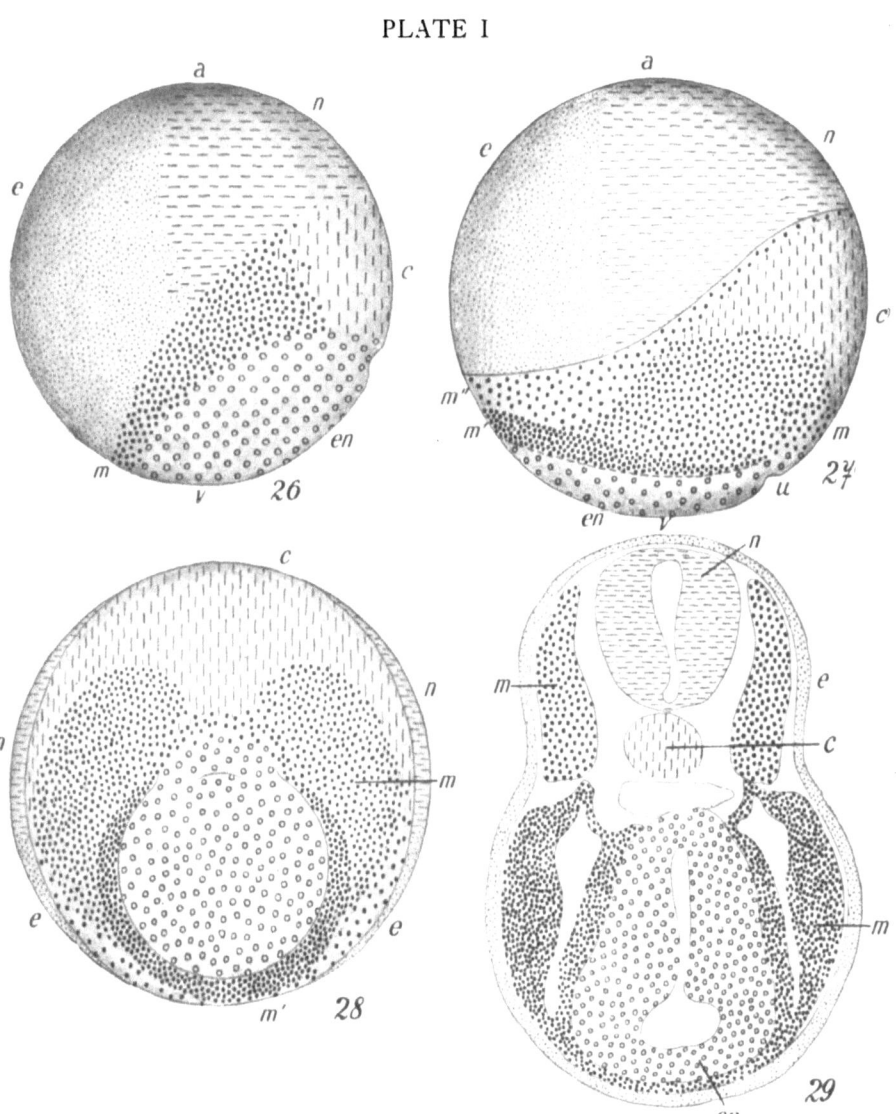

FIGS. 26 to 29.—The presumptive organ-regions in *Triton*: their topography in the germ and the positions they come to occupy in the embryo. (Semi-diagrammatic: based on the figures of Goerttler and W. Vogt). Fig. 26, blastula. Figs. 27 and 28, beginning of gastrulation. Fig. 29, transverse section of a young larva. Figs. 26 and 27 shew the germ from the left side; Fig. 28, from the vegetative or blastoporal side. The regions are marked in the same way as in Figs. 11 to 17. Ectoderm (epidermis), small dots; mesoderm, coarsely stippled; endoderm, small circles; neural (medullary) region, horizontal shading; notochordal region, vertical shading. *e*, ectoderm region (and epidermis); *en*, endoderm (gut); *m*, mesoderm of somites; *m'*, lateral-plate mesoderm; *m''*, mesoderm of tail-bud; *n*, neural (medullary) material; *c*, notochordal material; *u*, dorsal (cranial) lip of blastopore; *a*, animal pole; *v*, vegetative pole.

plane with the median plane of symmetry of the embryo; but very many exceptions are found. In the Urodela there is clearly no connexion between these two planes: the first cleavage divides the egg without reference to any symmetry relations which may later exist. The same holds good for the sea-urchin egg, as can be shewn clearly by *intra-vitam* staining. If, after completion of the first cleavage, the blastomeres are stained with Nile blue, larvæ are obtained each of which is half blue. But the plane which separates the stained from the unstained halves of the larva can make any possible angle with the symmetry-plane. Where cleavage is altogether irregular, as in the Insecta, there obviously can be no question of this relationship. The case of the cleavage plane is therefore analogous to that of the axis of the egg. Coincidence of that plane with a principal plane of the body may occur; but it is not necessary, and indeed it is not the rule—much less is there any predetermined obligation.

CHAPTER IV

THE POTENCY PROBLEM

I. THE POTENCY OF THE PARTS OF THE GERM

1. Regulative Eggs

(a) *Analysis of potency by redistribution of materials*

The demonstration that presumptive organ-regions have a definite topography means, in the first place, simply that definitely localized germinal materials pass into particular organs. It does not imply that this process is inevitable, particular organs being directly preformed in those germ-regions; neither can the conclusion be drawn that special regions of the egg or the germ are strictly limited to the production of one organ or tissue. As a matter of fact, the capabilities of these regions are more extensive than their actual performance in normal embryonic development. In other words: their potency is greater than their presumptive morphological value, or as it used to be put, their prospective potency is greater than their prospective fate. This is seen best in the so-called *regulative* eggs. These are eggs whose parts possess the ability to regulate their fate in development so as to produce a whole organism from one part; or in which the parts of the germ can yield a normal product as the result of an abnormal embryonic process.

The range of potency of individual germ-regions can be demonstrated in different ways. One way is by the redistribution of the germinal materials. If a particular part possesses only a limited potency, by virtue of which it can form one definite organ alone, then redistribution of the materials must result in a disturbance of development. Development must either come to a complete standstill, or it must produce an irregular confusion of tissues and organs, instead of an harmonious whole.

Redistribution of the germinal material can be brought about in many ways, of which cleavage of the egg under pressure is one.

If, for example, the eggs of *Paracentrotus* (an Echinoid) are compressed in the 4-cell stage between horizontal glass plates, in the direction of the primary egg-axis passing from animal to vegetative pole, the direction of the next cleavage is changed. As a result all the cells of the 8-cell stage lie side by side in one plane. This is contrary to the normal behaviour, according to which the third cleavage, which is horizontal, gives two superposed groups of four cells (Figs. 30 to 38). If the pressure is now removed a 16-cell stage arises which is in two tiers, for now the fourth cleavage-plane is a horizontal one. If the pressure is maintained longer, the 16-cell stage also is single-layered, all the divisions of the 8-cell stage then being vertical. The cells of the germ in this way are "shuffled." In the 8-cell stage the cells of animal and vegetative halves lie side by side in one plane, instead of one upon the other. But in spite of this a blastula is developed after removal of the pressure. This blastula differs from one produced by normal cleavage, in that the animal cells occupy a median circular band, which divides the vegetative cells into two parts. These two parts form the polar caps of the spherical blastula.

The egg of *Paracentrotus* is specially suitable for these experiments because in its vegetative half there is a ring-shaped zone of cytoplasm distinguished by its reddish pigment. This natural mark makes it very easy to follow the positions of particular regions during cleavage. The major part of this pigmented region normally lies in the vegetative half of the germ (Figs. 31, 32). Now, the normal 16-cell stage is produced by a meridional cleavage of the animal cells, and an unequal cleavage of the vegetative cells at right angles to this; there are thus divided off at the vegetative pole four small cells which do not participate in the red ring, and which go to form the skeleton. The four large vegetative macromeres, with the red ring, give rise to endo-mesoderm, on the other hand the eight animal micromeres yield ectoderm. Assuming that in cleavage

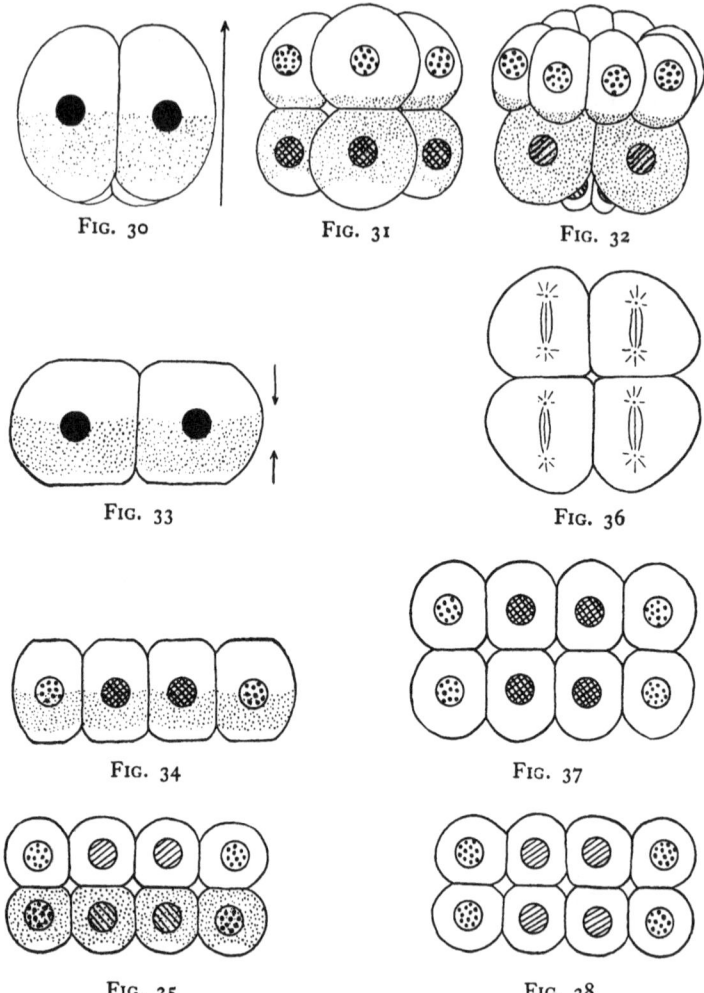

FIG. 30   FIG. 31   FIG. 32

FIG. 33   FIG. 36

FIG. 34   FIG. 37

FIG. 35   FIG. 38

FIGS. 30 to 38.—Diagrams shewing cleavage in the egg of *Paracentrotus*—
normal cleavage, Figs. 30 to 32; cleavage under pressure, Figs. 33 to 38.
Fig. 30, normal 4-cell stage side view; the direction of the egg-axis is given
by the arrow. Fig. 31, 8-cell stage. Fig. 32, 16-cell stage.
     Cleavage under pressure, side view—Figs. 33 to 35. Fig. 33, 4-cell stage
(direction of pressure shewn by arrows). Fig. 34, 8-cell stage. Fig. 35,
16-cell stage. Figs. 36, 37, and 38 shew the 4-cell, 8-cell, and 16-cell stages
respectively from the animal pole.
     Nuclei which correspond to one another are indicated by their identical
shading.

the potency of individual blastomeres may change, then by the third cleavage (Fig. 31) the endo-mesodermal cytoplasm and the endo-mesodermal nuclei would be separated from the animal or ectodermal cytoplasm, with its ectodermal nuclei. The next cleavage (Fig. 32) would divide off the skeletogenous region, with its special rudiments from the definitive endomesoderm. Upon this assumption, the whole gives the impression of a parcelling-out of pre-existent organ-regions possessing a partial but definite potency.

In the case of cleavage under pressure something very different takes place. In the third cleavage (Fig. 34) *all* the cells receive "endo-mesodermal" cytoplasm, recognizable by its red pigment-ring. Nuclear division, as such, is not affected by this. That is to say that four of the cells, with identical cytoplasm, receive ectodermal nuclei, while four others receive endo-mesodermal nuclei. If, after releasing the pressure, the fourth cleavage now occurs horizontally, the endo-mesodermal cytoplasm is separated, it is true, from the ectodermal (Fig. 35); but no corresponding separation of the nuclei is possible, since in the previous division the endo-mesodermal nuclear material was already separated from the ectodermal. Therefore in the 16-cell stage, four cells which by virtue of their cytoplasm have the value of endo-mesoderm, receive ectodermal nuclei; four other cells possess endo-mesodermal nuclei; four nuclei, however, are already skeleton-forming, since these are separated from the vegetative germ material in the fourth division. Obviously the four other remaining cells, which are present, combine ectodermal cytoplasm with ectodermal nuclear material. The fifth division, which now follows, cuts off the clear vegetative caps (mesenchyme) to form eight cells (octet). Of these, four must contain ectodermal nuclei, and only four possess normal nuclear material. The archenteron (endo-mesoderm) is formed from cells which at this time still contain the red pigment-ring. After the separation of the mesenchyme there is found in one portion of the presumptive archenteron-cells nuclear material which is partly ectodermal, partly mesenchymal. All these relations can be best understood by reference to Figs. 30 to 38, where the difference in character of the nuclei is indicated by stippling, shading, etc. In any case, when cleavage takes place under pressure, the material of the germ undergoes a complete redistribution. Cells of the animal hemisphere contain not only ectodermal but in part endo-mesodermal nuclear material; those of the vegetative half—with endo-mesodermal cytoplasm—receive partly ectodermal nuclei; archenteron cells contain partly ectodermal, partly mesenchymal nuclei; while mesenchyme, in addition to its appropriate nuclei, has ectodermal nuclear material. If the hypothesis of a parcelling-out of germinal materials with limited potentialities be correct, then it is clearly impossible for a normal animal to arise from such a germ. But in actual fact entirely normal larvæ are developed after the pressure has been removed. Therefore in the pressure experiments the embryonic material must have been put to a use different from that which it has in normal development. If so, its capability (potency) is different from—and greater than—its actual performance in normal ontogeny.

Redistribution of the germinal materials can be even more strikingly demonstrated by mechanical displacement of the first cleavage-cells or blastomeres, or by artificially uniting the blastomeres of different germs to form a new starting-point in development. Such experiments may be explained by means of the diagrams in Figs 39 to 43.

The first two cleavages of the newt's egg are meridional: seen from the animal pole, the germ appears as four cells, of which two are always derived from one and the same $\frac{1}{2}$-blastomere of the 2-cell stage (Fig. 39). In the course of the first cleavage the first two blastomeres pass through a stage of maximum separation from one another, when they are united only by a cytoplasmic bridge, so that the egg is dumb-bell shaped. Later on, this bridge too is broken down, the two blastomeres then lying with their inner surfaces close together. The dumb-bell formation is particularly marked if the vitelline membrane has first been removed (Fig. 40). If the first two blastomeres are artificially separated from one another, then each during the next cleavage shows a similar dumb-bell stage, which now consists of the $\frac{1}{4}$-blastomeres still united by a cytoplasmic bridge (Fig. 41). This shape assumed by the two

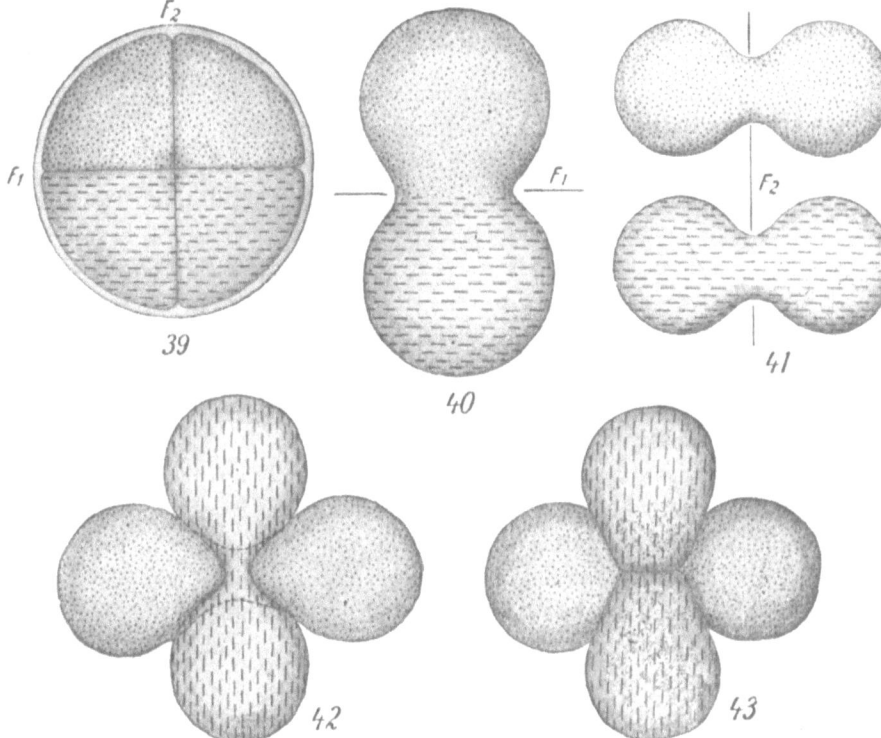

FIGS. 39 to 43.—Diagrams illustrating the experimental rearrangement of the blastomeres of *Triton*. Fig. 39, appearance of the normal 4-cell stage from the animal pole. Fig. 40, "dumb-bell" phase during the first cleavage of an egg freed from its vitelline membrane. Fig. 41, dumb-bell phase of the two artificially separated $\frac{1}{2}$-blastomeres in the course of the second cleavage. Fig. 42, the two dumb-bell-shaped $\frac{1}{2}$-blastomeres laid one across the other. Fig. 43, the rearranged germ seen from the animal pole after completion of the second cleavage. The kind of rearrangement produced can be appreciated by comparing this with Fig. 39: $\frac{1}{4}$-blastomeres derived from the same $\frac{1}{2}$-blastomere are indicated by lines or dots as the case may be. F_1, first cleavage-plane; F_2, second cleavage-plane.

halves of the germ makes it possible to lay one half crosswise upon the other (Fig. 42). After the completion of the cleavage actually in progress, the superposed blastomeres sink between the underlying pair, and in this way there is formed a germ which has, exactly like a normal one, four cells, but whose blastomeres obviously are in a different spatial relation from those arising in normal cleavage. This will be understood without difficulty by reference to Figs. 39 and 43. Normal embryos may arise from germs which have been disarranged in this way.

The result, however, is much more striking if, instead of simply redistributing the blastomeres, we produce a germ out of four ½-blastomeres by a procedure corresponding to that which we have described. Like a normal germ after two cleavages, it will have four cells, but each of its cells will have the value of a ½-blastomere. In such an experiment the two pairs of ½-blastomeres can be taken each from a different species, e.g. from *Triton tæniatus* and *T. alpestris* (Pl. II, Fig. 44). From such a heterogeneous fusion of germs there may develop normal embryos, which are only remarkable for their size. An organism is thus produced which has two fathers and two mothers, since it arises from two normally fertilized eggs. In the case of all such experiments the parts of the germ are forced to take a path in development other than that which is followed in the normal course of events. This illustrates particularly well the fact that the potency of blastomeres may be different from their actual performance in development. It would indeed be hard to find a better proof of the epigenetic plasticity of development.

(b) Determination of potency by isolation of blastomeres

In their normal relations of contact in the germ, the individual blastomeres have a definite role in development, since they produce a definite part of the new organism. That there resides in them a greater potency is shewn, however, by freeing them from this contact and allowing them to stand, so to speak, on their own feet. If the first two blastomeres of the Newt's egg be completely isolated from one another, a single egg gives rise to two normal larvæ whose only difference from the larva produced by a whole egg is that they are smaller. Such duplicities of development can be produced fairly easily

PLATE II

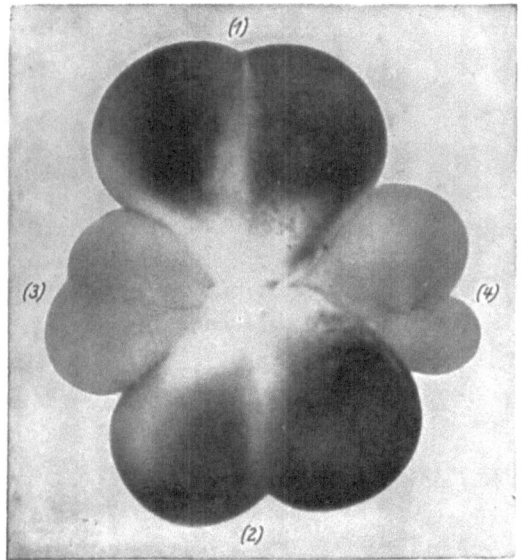

FIG. 44.—Regular criss-cross fusion of the germs of *Triton alpestris* and *T. tæniatus* to form a single germ —brought about by laying them crosswise over one another in the early 2-cell stage. The four ½-blastomeres have settled down into one plane, and their next division is already well advanced. Sectors 1 and 2 are derived from *T. alpestris*; sectors 3 and 4 from *T. tæniatus*. (After Mangold and Seidel.)

if, during the formation of the first cleavage-furrow, the egg, still within its envelopes, be constricted by means of a noose of hair, so that the first two blastomeres are completely separated from one another (Fig. 45, A). After a time the noose is loosened somewhat, and two normal larvæ are formed, whose origin from a single egg is sufficiently manifest from the fact that they

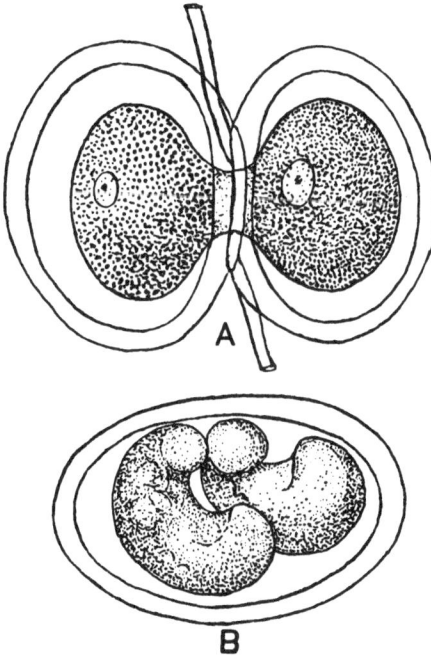

FIG. 45.—The production of a double structure from the egg of *Triton tæniatus*. *A* dumb-bell stage in the separation of the two ½-blastomeres by means of a noose of hair; *B*, the resulting twin embryos in the same egg-envelopes. (From Korschelt, 1927, after Spemann.)

are in the same envelope (Fig. 45, *B*). Incomplete constriction may lead to partial twinning.

The egg of Amphioxus is also a typical regulative egg. Isolated ½- and ¼-blastomeres may develop so far beyond the embryonic stage as to produce fully formed young larvæ (Fig. 46). This alone makes it clear that the different regions of the egg must here follow a course in development differing

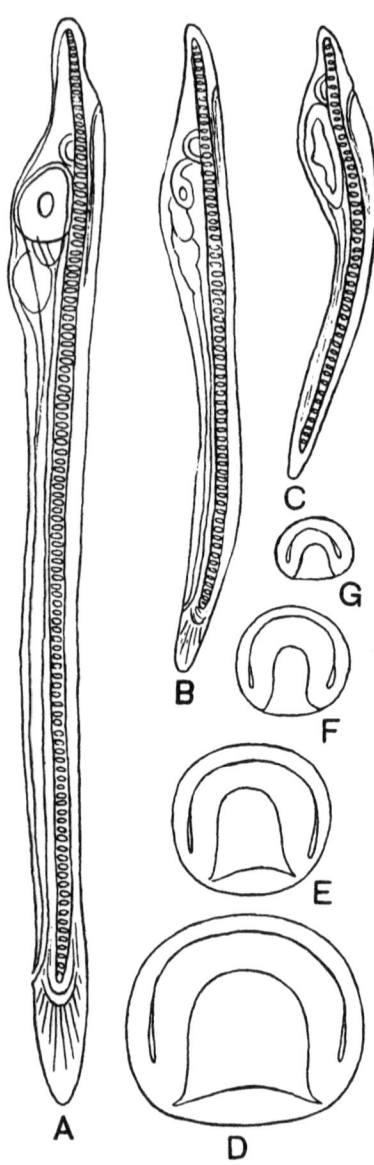

Fig. 46.—*A–G*, whole larvæ from isolated blastomeres
of *Amphioxus lanceolatus*. *A*, normal young larva
(first gill-slit only present); neural tube, notochord,
gut, and cœlomic sacs clearly visible. *B*, young
larva from an isolated ½- (or ⅔-) blastomere; *C*,
from an isolated ¼-blastomere; *D*, normal gastrula;
E, *F*, *G*, gastrulæ from ½, ¼, and ⅛-blastomeres
respectively. (From Korschelt, after Wilson.)

considerably from the normal one, thus revealing their wider range of potency. When a $\frac{1}{8}$-blastomere of Amphioxus is taken we find that it is now no longer possible to obtain a larva, but this fact does not affect the point at issue; gastrulation of a germ derived from a $\frac{1}{8}$-blastomere occurs with some difficulty, though, indeed, in many cases gastrulæ are formed. The animals developed from isolated blastomeres differ from the normal ones only by being smaller.

Complete larvæ can also be raised from the $\frac{1}{2}$-blastomere and even from the $\frac{1}{4}$-blastomere of the sea-urchin egg, and to this regulative type belong also the eggs of Cœlenterates, Nemerteans, Cyclostomes, and Teleosts. It must be observed, however, that gradations of behaviour are found between the individual groups, so that the power of regulation possessed by the parts is by no means equally obvious in all. Where it is very slightly marked in the case of individual blastomeres, we speak of *mosaic* eggs.

2. Mosaic Eggs

In typical regulative eggs the limits of potency of individual regions of the young germ are manifestly greater than the normal performance of these parts. The so-called mosaic eggs give—at least at first sight—a very different impression. In them the isolated $\frac{1}{2}$-blastomere generally does not produce a whole embryo at all; indeed in extreme cases it produces only a half. But in any case these embryos from separate blastomeres show substantial defects. This is especially true of those eggs in which particular regions can be identified in the cytoplasm before cleavage, as we have seen above. In extreme cases the course of development is established or determined before fertilization, at all events before cleavage. Thus the resulting blastomeres vary greatly in their capacity, since they have already been specialized for definite functions. This can often be recognized, indeed, from visible differences in character due to their very unlike differentiation. As a consequence of this initial determination and differentiation, the blastomeres, in following the course marked out for them in development, shew considerable independence and individuality in their further development; and at the same time a limitation of their developmental capabilities. Development, in this way, comes to be a mosaic: the individual germ-regions, with their exclusive determination, contribute to the complete organism independently of one another, just as the little stones in a mosaic picture, themselves independent units, combine to form the whole. If a part of the mosaic of germinal regions is removed, the structure is incomplete; nor can it become a whole by processes of regulation—hence the term mosaic eggs.

Apart from the very complete differentiation of these eggs before cleavage, a general characteristic of them all is their rapid progress through the first developmental processes. As one would suppose *a priori*, both the mosaic character of development and the lack of a capacity for regulation are exhibited to a varying extent; there is, indeed, never a complete absence of regulatory activity after interference with normal development, or after removing parts of the very young germ. Many cases, however, exist in which regulative processes are so slight as hardly to be recognizable.

The eggs of Ctenophores, Nematodes, Annelids, Molluscs, Arthropods, and Ascidians are generally reckoned as mosaic eggs.

We have already dealt fully with the determinative character of cleavage in the eggs of Nematodes and Ascidians. Here we may consider more closely the behaviour of the molluscan egg, in connexion with what is known of the egg of *Dentalium*. The cleavage of the egg of the mollusc is peculiar in that a special part of the cytoplasm, the "yolk-lobe," or polar lobe, is constricted off from one of the first four blastomeres. In the egg of *Dentalium*, before cleavage, three zones can be recognized—a broad pigmented belt, and two clear polar caps. During cleavage the clear cap at the vegetative pole passes into the polar lobe, which is then annexed by one of the first two blastomeres—the so-called *CD* cell (Figs. 47–52). In the second cleavage this polar lobe again separates distinctly from

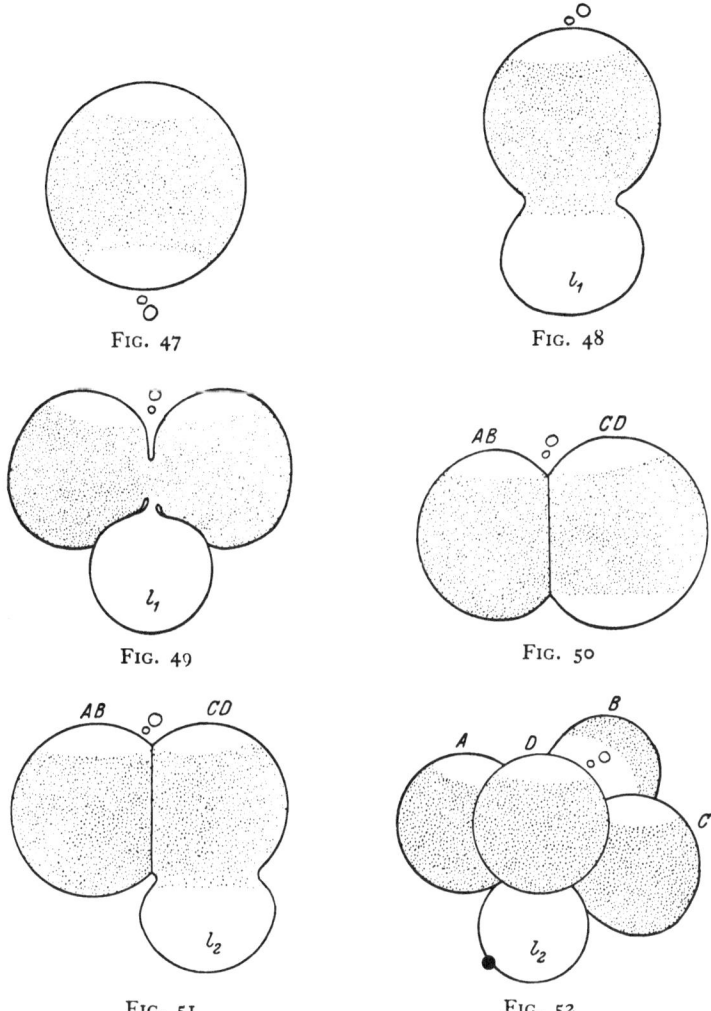

FIG. 47

FIG. 48

FIG. 49

FIG. 50

FIG. 51

FIG. 52

FIGS. 47 to 52.—Cleavage of the egg of *Dentalium*. The ¼-blastomeres are lettered A to D. l_1 and l_2, first and second yolk-lobes respectively, which ultimately become part of blastomere D. Fig. 47, egg before cleavage; pigment-band stippled. Fig. 48, beginning of first cleavage. Figs. 49 and 50, 2-cell stage; Fig. 51, beginning of second cleavage; Fig. 52, 4-cell stage. (After Wilson.)

the cytoplasm above (Fig. 51), and passes completely into one of the ¼-blastomeres—the so-called *D* cell. The mosaic character of the cleavage is shewn in this behaviour. If one removes the first polar lobe, formed in the early stages of the first cleavage, the following division is symmetrical, but the resulting larva now lacks definite organ-regions—the apical organ and the post-trochal region. If the second yolk-lobe, which is

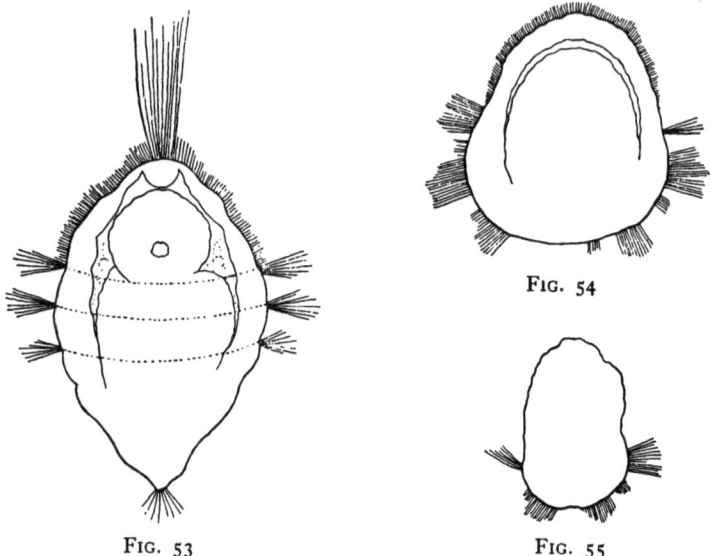

FIG. 54

FIG. 53   FIG. 55

FIG. 53.—Normal Trochophore larva of *Dentalium*, 24 hours old. Fig. 54, a 24-hour larva of the same which developed after removing the primary yolk-lobe—post-trochal region and apical tuft of cilia missing. Fig. 55, defective larva of *Dentalium*, 72 hours old, developed from one of the small blastomeres *A*, *B*, *C* not containing the yolk-lobe.

formed during the second cleavage from the *CD* cell, is removed, the larva will then possess the apical organ, but it will lack the post-trochal region (Figs. 53, 54). These defective larvæ cannot metamorphose; they are completely arrested in development at a rudimentary stage. From the isolated *AB* half of the egg (Fig. 50), or from the isolated ¼-blastomeres *A*, *B*, *C*, very defective larvæ always arise, because the cells in question have received none of the substance of the yolk-lobe (Fig. 55). From the isolated half *CD*, or from the ¼-blastomere D, with which the yolk-lobe finally fuses, a larva is developed which possesses all the organ-rudiments; these larvæ also achieve metamorphosis as would be expected.

In any case, it is impossible to produce twinning by simply isolating the two $\frac{1}{2}$-blastomeres of the *Dentalium* egg, as we have found was possible in other eggs. Now, it happens in abnormal circumstances that the polar cap is not completely united with the D cell, but is shared instead between two cells. As a result, the germ possesses, as it were, two D cells. If now such a germ be divided into its two halves, a whole germ is produced from each half; and in these conditions the egg of *Dentalium* gives rise to twinning. The development of these twin larvæ has not been followed very far, but their character shews that, even here, one part of the egg is capable of producing more than it does in normal development.

Researches on the eggs of other molluscs have similarly shewn the predominantly mosaic character of their development. The egg of the worm *Tubifex* behaves in a similar way. It contains a special *pole-plasm* corresponding to the yolk-lobe. This pole-plasm is normally allotted to a certain $\frac{1}{4}$-blastomere, but where special means are taken to distribute it between two cells, the egg of *Tubifex* also can form twins.

The Ctenophore egg is often taken as an example of a mosaic egg (*Beroë ovata*). The isolated halves of the egg produce only defective larvæ, each of the two partial larvæ lacking, in particular, some of the characteristic rows of combs. As is well known *Beroë* possesses, like all Ctenophores, eight rows of swimming-combs, which, as the so-called ribs pass round the body of the animal from the sense-plate downwards. Now, the partial larvæ have, instead of eight, only four of such ribs. At the same time there is by no means a complete absence of regulation. For example, each partial larva possesses a complete statocyst and a gut, though the latter may shew local defects. The partial larvæ are at any rate not true half larvæ.

In some Arthropods the eggs belong very definitely to the mosaic type, while in others considerable powers of regulation are shewn by the parts. For example, isolated blastomeres of the egg of *Cyclops* produce only extremely defective larvæ. In other Arthropods it is quite possible to produce experimentally double and multiple monsters, thus proving that the potentialities of the parts concerned are not confined to their normal developmental performance. This is true, for instance, of the egg of the dragon-fly, *Platycnemis pennipes*. If the hinder pole is carefully cauterized early in development, before visible

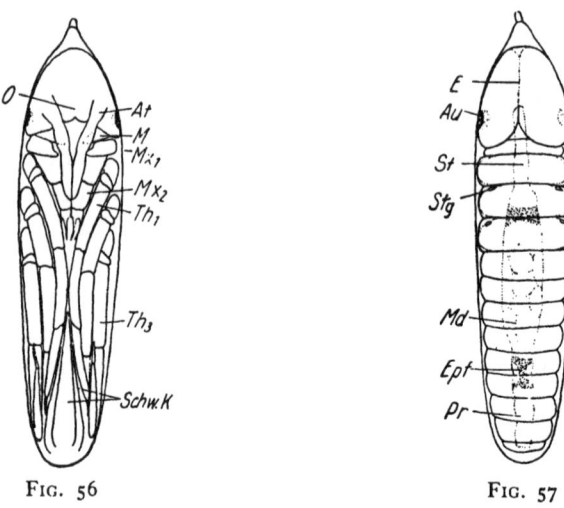

FIG. 56   FIG. 57

FIGS. 56 and 57.—Normal embryo of *Platycnemis pennipes* during the period of histological differentiation. Fig. 56, ventral aspect. *At*, antenna; *M*, mandible; *Mx₁*, *Mx₂*, maxillæ; *Th₁*, *Th₃*, thoracic limbs; *O*, labrum; *Schwk*, caudal gills, folded forward from their bases towards the anterior pole of the egg. Fig. 57, dorsal aspect. *E*, indentation of head-capsule; *Au*, eye; *St*, stomodæum; *Stg*, stigma; *Md*, mid-gut; *Epf*, entrance to hind gut; *Pr*, proctodæum. (After Seidel.)

differentiation of the germinal rudiment, there arise very fine, often scarcely visible, longitudinal folds in the surface of the egg due to an unequal stretching of its materials by the unequal heating. This produces the same effect as an incomplete constriction of the material of the germ. Were the development strictly mosaic in type we should now expect two somewhat defective half-structures to be found at the hinder end. This, however, does not occur; on the contrary a real duplicity is developed; that is to say considerably more is formed from each half of the germinal material than it normally gives rise to. A normal embryo of *Platycnemis* during the period of histological differentiation shews clearly in ventral aspect, through its egg membrane, the tail-processes, appendages, and mouthparts (Fig. 56). In the dorsal aspect it is the gut which is most plainly seen (Fig. 57). After cauterization of the hinder end in the blastoderm stage, this end of the embryo is widened; and instead of the three tail processes normally present, six

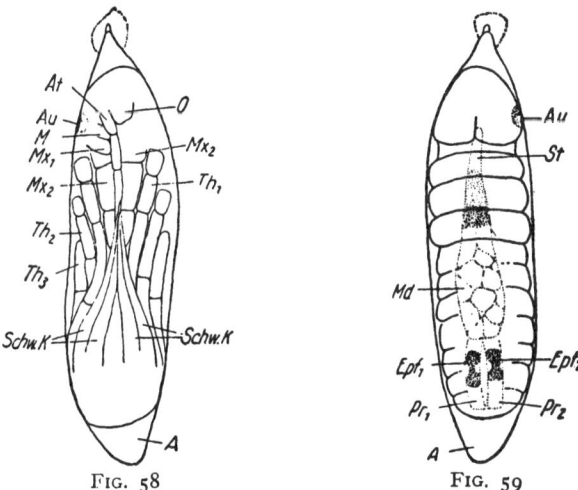

FIG. 58   FIG. 59

FIGS. 58 and 59.—Embryo of *Platycnemis pennipes* with experimentally produced doubling of the posterior end. Fig. 58, ventral aspect: the left eye, left antenna, and apparently also the left mandible and left first maxilla are absent; caudal gills doubled. *A*, deficiency due to cauterization. Fig. 59, dorsal aspect shewing duplicity of hind-gut. Epf_1 and Epf_2, entrances to the two hind-guts; Pr_1 and Pr_2, the two proctodæa; the rest of the lettering as in Figs. 56 and 57. (After Seidel.)

now arise. Only four of these are seen in Fig. 58, since on either side one of them is hidden. The duplication of the hind-gut as seen in dorsal aspect is plainly shewn in Fig. 59. The mid-gut (*Md*) is single, but there follow two hind-guts (Pr_1 and Pr_2) with their two entrances (Epf_1 and Epf_2). Each of the hind-guts has its own anus under the proximal part of one of the groups of tail-processes. As a result of the cauterization of the hind end a defect is produced which leads to the space being diminished, so that the tail-processes and a part of the abdomen are bent towards the ventral side, and the former, though they are no longer than in a normal embryo, reach farther forward. Further study shews that the hinder end of the ventral nerve cord is also involved in the duplicity.

After what has been said above about the differentiation of presumptive organ-regions, it is not surprising to find that the egg of the Ascidian shews a pronounced mosaic character in its development. After isolating the blastomeres there is never regulation to form a whole, whether in the case of $\frac{1}{2}$-, $\frac{1}{4}$-, or $\frac{1}{8}$-blastomeres; in every case a defective larva is formed. Though the larva produced by a $\frac{1}{2}$-blastomere may be covered with ectoderm, and so give the impression that it is complete, yet the paired organs are always developed only on one side. The axial organs are of course not strictly mere half-structures; nevertheless the potentialities manifested by the cleavage-cells of a germ at this stage do not in any essential way go beyond the actual performance in development of these cells.

II. Potency in Regenerative Processes

1. Regeneration as a Supplementary Developmental Performance

The whole of the problem of potency has by no means been accounted for when the potency of the blastomeres has been defined. This is clear from the fact that not only blastomeres but often parts of older embryos with the rudiments of their adult organs already present can shew a range of potency in excess of their presumptive morphological value. This wider range of potency explains the fact that the fate of these parts in advanced stages is frequently indeterminate. To a certain extent this indeterminateness shews itself even in individual organ-rudiments. Naturally in this respect the behaviour of different kinds of animals varies, and the question as to whether a given egg should be considered as belonging more to the regulative or more to the mosaic type becomes important. We will not discuss here the range of potency possessed by advanced stages of development and by the rudiments of organs: later, when dealing with the problem of determination, we shall have occasion to recur to that matter. Let us first consider potency in relation to those secondary developmental processes which are seen in regeneration.

The mere fact of regeneration proves that power to develop has not been lost by the fully developed adult organism. What, however, commands our special interest is the fact that regeneration, as opposed to the primary developmental processes of ontogeny, always involves a supplementary performance—a *supererogation*.[1] Let us take the simple case of an organ which, after having been formed in ontogeny, is lost and then formed again. Normally the organ is in relation to a particular presumptive organ-region which actually develops into the final product. After the removal of the completed organ this germinal region is no longer available as a starting-point; there must therefore be another source of origin for the part regenerated. In this way the supplementary character of what is achieved in regeneration shews that there exists really a greater range of potency than was apparent in the first formation of the organ. This will perhaps be seen more clearly when the origin of the material for regeneration and the questions of metaplasia and heteromorphosis have been discussed.

As to the simple fact of this supererogation, it is specially noteworthy that the power of regeneration is not limited to organisms developed from regulative eggs, but is found also in those derived from typical mosaic eggs. We have seen that when single blastomeres are removed from developing eggs of the latter kind defective development results, and that true regulation to form a complete embryo does not generally occur. If, then, animals of this particular type possess later the ability to regenerate lost parts, their incapability of regulation during ontogeny cannot rest upon a behaviour of the organ-rudiments which is fundamentally different from that in regulative eggs, but must be due to the intervention of special processes which diminish or check the potency of the separate parts during ontogeny.

[1] *Mehrleistung*: see Translators' Preface.

Now, some forms of this kind shew an extraordinarily limited power of regeneration; Nematodes, for example, can make only an attempt at regeneration. There are other forms, however, in which regenerative ability is very great, as in the Ascidians. The case of *Clavelina lepadiformis* is well known in this connexion. The body of the animal is two or three centimetres long, and we can distinguish in it four regions (Fig. 60). A middle, narrow region (II), containing the fore- and hind-gut

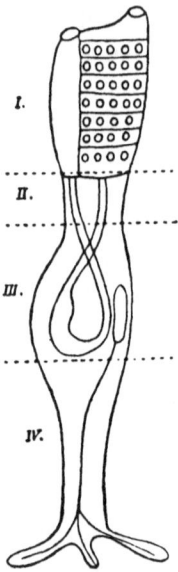

Fig. 60.—Diagram shewing the regions (I–IV) of the body of *Clavelina lepadiformis*. (After Driesch.)

(œsophagus and rectum), unites the so-called pharynx (I) with the true visceral sac (III). The pharynx, as in all Ascidians, is simply the first part of the gut, perforated by numerous openings—the gill slits—lined with ciliated cells, and thus forming a fine meshwork that functions as a gill. To this part of the body also belong the two great openings for the incurrent and excurrent water, and between these the large cerebral ganglion. In the visceral sac (III) are the stomach, intestine, heart, reproductive organs, etc. This is followed by the region of the stolon (IV), in which no organs of the body are found.

If the pharynx of a full-grown *Clavelina* be removed by a transverse cut, both this and the remaining part of the body produce at the cut-surface a regeneration-blastema from which the missing parts are re-formed. This process has a special importance for us because the pharyngeal region regenerates organs, such as the heart and the reproductive organs, of which it contained no trace before. The same is true of the regeneration of a pharyngeal region by the rest of the body: gill-slits, nerve centres, etc., are replaced, though no parts of these existed in the stump. The supererogatory character of regeneration is here clearly seen. The source of the regenerated organs is different from what it was in their ontogeny, in which their rudiments were only to be found in special cells; it was these cells that developed into the organs which were cut off, and they are therefore no longer available as points of origin for the regeneration of those organs. Thus it is seen that regions which in ontogeny have an absolutely definite and exclusive destiny, can nevertheless perform more than would appear from consideration of their embryonic development alone. All Ascidians do not, however, possess the same capacity for regeneration. In *Ciona*, for example, it seems to be limited to that region of the body which surrounds the visceral sac. But that is not of fundamental importance in this connexion.

The fact that potency is not completely exhausted during ontogeny is particularly clear in the cases which occur of multiple or super-regeneration and of repeated regeneration. Observations, for example, on worms and on the limbs of Urodeles shew that a particular region of the body can be repeatedly regenerated. Thus fourteen successive regenerations of the tail end of *Lumbriculus* are recorded. Regenerated parts are themselves also capable of regeneration. We may speak of super-regeneration when an organ, without being removed, is formed a second time by regenerative processes; or when a particular organ arises simultaneously in multiple form by regeneration. In this way we get double and multiple structures, such as are found in all groups of animals which have any power of regeneration at all. Not only does this occur frequently in nature, but it can be produced experimentally. Doubling and multiplication not only of the appendages of the body (tail, extremities, etc.), but also of whole segments of the body arise in this way. Such phenomena are met with not only in animals whose eggs clearly belong to the regulative type, but also in

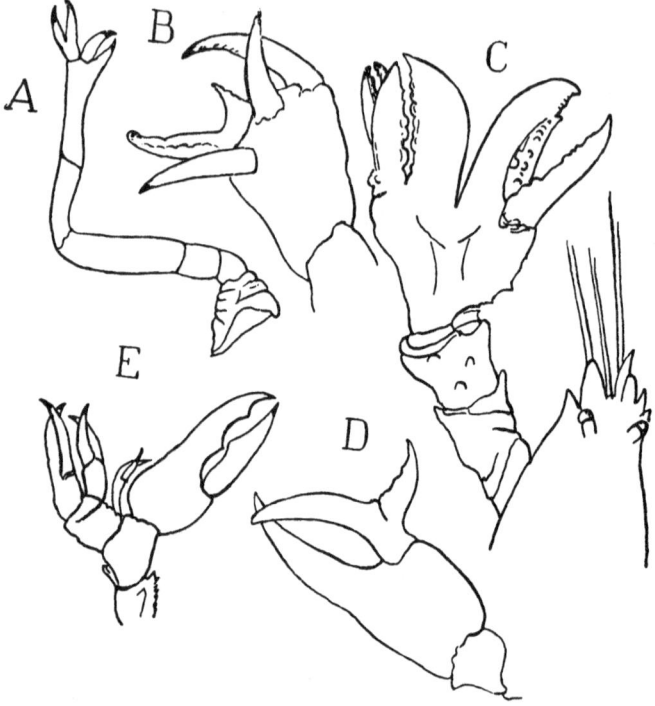

FIG. 61.—Double and multiple structures in the appendages of Crustacea. *A, Cambarus bartoni,* third left walking-limb; *B, Platycarcinus pagurus,* right claw; *C, Homarus americanus,* left claw; *D* and *E, Astacus fluviatilis,* right and left claws respectively. (From Korschelt, after Przibram.)

those whose development is in the main mosaic in character. Thus, multiple regenerates are not at all infrequent among the Arthropoda, the eggs of which tend, though not always in an extreme degree, towards the mosaic type. Fig. 61 shews examples of this in the doubling or trebling of the claw of the crab. This last example may be regarded as a triple structure due to fracture, in which after injury to the organ originally present there arises beside this a double regenerate. Triple structures of this kind occur in the most varied animal types.

The simultaneous occurrence of multiple regenerates depends largely upon the condition of the wound-surface from which regeneration proceeds. It has been shewn that the orientation of a regenerate is not decided by the axis of the body or by that of the organ, but by the orientation of the wound-surface. In general the regenerate at first stands at right angles to the wound-surface, and it would appear *a priori* that, given a complicated wound, the several parts of whose surface were inclined to one another, there might be formed several centres of growth, and hence several regenerates

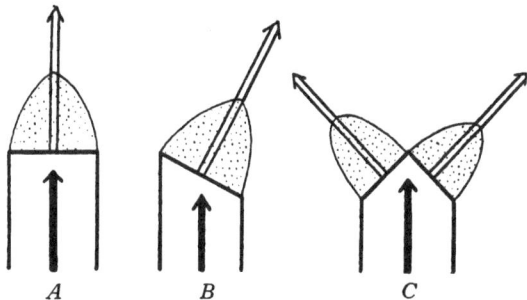

FIG. 62.—Diagrams illustrating the relation of the wound-surface to the orientation of the regenerated part. *A*, axis of regenerate (white arrow) coincident with original axis (black arrow); *B*, regenerate forming an angle with original axis; *C*, double regenerate arising from "roof-shaped" wound-surface.

(Fig. 62). This has been experimentally confirmed by making appropriate amputations. The special symmetry-relations often existing among the regenerates in such cases need not concern us here.

2. Origin of the Material of the Regenerated Part. Metaplasia

If the origin of the material out of which a regenerate is formed is investigated, the supererogatory character of regenerative as opposed to embryonic development appears very clearly in many cases, and we obtain further confirmation of the fact that the limits of the potency of a particular region go beyond the structure that is actually formed.

In regeneration it by no means always holds that "like begets like." It is true that regenerated tissue is formed not infrequently from tissue of the same sort which has remained behind, but often it is otherwise; and indeed from this point of view regeneration may be a contradiction of ontogeny. A germ-layer which is generally defined precisely in ontogeny is by no means always thus limited in regeneration. For example, in the regeneration of the appendages of Decapods and Isopods the new musculature arises not from the muscles of the stump but from material derived from the hypodermis; in fact, the muscles in the regenerate are of ectodermal, not mesodermal, origin. Small pieces of Nemerteans, cut from in front of the mouth-opening, contain none of the gut. They are able to develop into whole animals, and to re-form the missing organs. Thus the gut obviously cannot here be of endodermal origin. It is formed in fact from the cells of the wall of the rhynchocoel, with the addition of numerous cells from the surrounding parenchyma, which are undoubtedly mesodermal in character. This is a characteristic case, which throws light on the potentialities of the cells in question.

The regenerated part has also a foreign origin when it has to grow altogether unconnected with any remainder of the original organ. Something of this sort can be brought about experimentally in the regeneration of the limb of the newt (*Triton*). A completely exarticulated bone of such a limb is not regenerated. But if the humerus is carefully removed and the limb is now amputated in the region of the upper arm, from which the bone has been taken, there follows a regenerative new growth upon the bone-free surface. The regenerate, in spite of this, possesses all the typical parts of its skeleton, which thus must arise not from a remainder of similar tissue but from cells of the regeneration-blastema. Figs. 63 and 64, Pl. III, are radiographs shewing experiments on *Triton cristatus* in which the regenerates are well advanced in growth. After the removal of the scapula and the humerus the fore-limb was

PLATE III

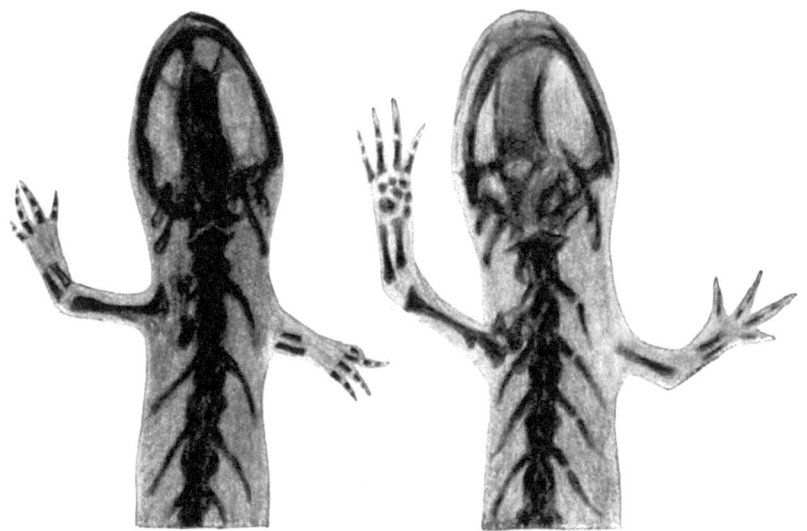

FIGS. 63 and 64.—*Triton cristatus;* regeneration of limb-skeleton from a stump containing no bone. Fig. 63, amputation above the elbow after removal of scapula and humerus; regeneration of arm and hand; carpus still cartilaginous, but radius and ulna (displaced into bone-free upper arm) and phalanges already ossified. Fig. 64, amputation of the limb at its base after removal of scapula and humerus; hand, fore-arm, and upper arm regenerated, including skeleton; the scapula and the proximal part of the humerus are missing. (From Korschelt, after P. Weiss.)

amputated in the one case (Fig. 63) just above the elbow, and in the other case (Fig. 64) close to the body. In both cases the distal parts of the limbs, together with the skeleton, were re-generated, and there was no possibility that derivatives of the bony tissue played any part in the formation of the blastema. Other material must therefore have been used, which could only be possible as the result of an adequate potency in those cells which contributed to the blastema.

The cells which form the basis of the regenerate are in most cases what may be called "indifferent," that is they are not the same as the specifically differentiated tissue cells found in the neighbourhood of the place of amputation. It has been estab-lished in the case of Annelids that the formation of the re-generate depends on *neoblasts*. These are large amœboid cells, differing from the other cells of the body in their structure and size. They are widely distributed throughout the body of the worm, though they have not been demonstrated every-where, and they wander, often from a great distance, to the place of regeneration. The exact manner of their appearance varies greatly in different species. In view of the fact that the Annelid egg undergoes cleavage of a strongly determinate character, the appearance of such distinctly pluripotent cells is very interesting; for in mosaic eggs at an early stage only unipotent blastomeres are apparently present. The same is true of Ascidians, where wandering cells take part in the formation of the regenerate.

A particularly impressive illustration of the range of potency in regeneration is furnished by cases where the regenerate is not developed from these "indifferent" cells, but its starting-point consists of the fully differentiated cells of a normal tissue or organ of the body. In this way these cells produce tissues or organs completely foreign to them, and with which they are otherwise unconnected. Not only, then, is the presumptive course of development diverted into other channels, but the normal fate in development, which has already been fulfilled, is set aside by this final differentiation, and a new, secondary developmental product substituted. The way in which the range of potency surpasses what is normally done in development could hardly be more plainly shewn.

This transformation of one kind of tissue into another is generally called *metaplasia*. In this process it is rare (though not altogether unknown) for one definite kind of tissue-cell to be directly responsible for the formation of the cells of another specific tissue. It is, however, much more generally the case in metaplasia for a particular cell first to lose its previously acquired specific nature, to become again capable of division, and then to produce new cells, which finally assume a differentiation other than that which they originally possessed. In the case of regeneration from dissimilar material such metaplasias necessarily occur, as in the formation of muscles from ectodermal cells of the hypoderm of the crab's limb, and in regeneration of the bony skeleton from the bone-free stump of a limb.

A specially interesting case of metaplasia is that of the regeneration of the eye in urodele Amphibia (Figs. 65–67). If the lens is removed from the eye of *Triton* (this can be done by a smooth transverse cut above the pupil, followed by slight pressure on the eyeball), it is completely re-formed in a few weeks. In young animals regeneration is more rapid. The lens during embryonic development is formed from the epidermis of the head of the embryo. Regeneration, however, does not start from this normal source—represented in the fully formed eye by the conjunctiva and the corneal epithelium—but from the iris, which is derived from the rim of the primary optic vesicle. On the upper edge of the pupil a vesicle is formed by separation of the two epithelial layers of the iris. This is the rudiment of the lens (Fig. 65). The further development of the lens is carried out exactly in the embryonic fashion. The vesicle is finally cut off completely from the iris (Fig. 66), its inner wall thickens and forms the lenticular nucleus, and the substituent lens has in the end exactly the same structure as the normal organ. Thus not only does the regenerate come from a foreign source, but iris-cells, specifically differentiated as such, turn into lens-cells and crystalline fibres. The capacity of the specialized cells of the eye of *Triton* is greater still. If lens and

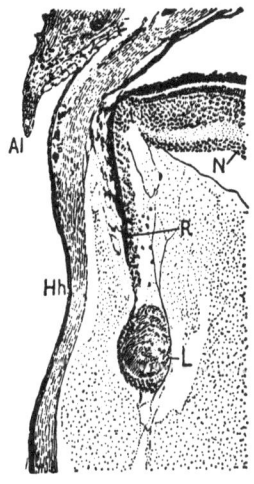

FIG. 65

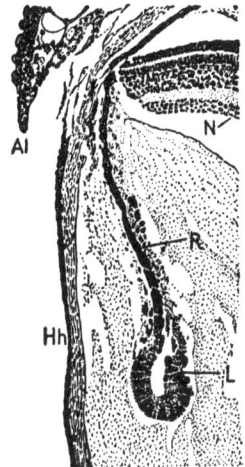

FIG. 66

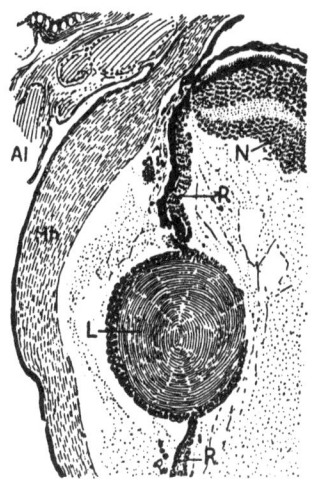

FIG. 67

FIGS. 65 to 67.—Regeneration of the lens in the eye of *Triton*. Fig. 65, first rudiment of the new lens-vesicle formed from the iris at the upper edge of the pupil. Fig. 66, the lens begins to free itself; the inner wall of the lens-vesicle is greatly thickened, and differentiated as a lens "nucleus." Fig. 67, substituent lens almost completed. *Al*, upper eyelid; *Hh*, cornea; *L*, lens-vesicle, and lens; *N*, retina; *R*, iris.

retina are removed from the eye of the larval *Triton tæniatus*, by an operation on the inner side of the eye, a great part of the vitrous humour being lost in the process, the eye is completely re-established in a short time. Regeneration of the lens is completed, as already described, from the upper edge of the unremoved iris. The removal of the retina has left this with a damaged inner edge; from this arises the material for the new retina, which also, however, draws upon the persistent *tapetum nigrum* for material. In this latter process the tapetal cells lose their pigment and proliferate—another good example of the conversion of what is normally a final product of development into a secondary one which is specifically different from it.

3. Heteromorphosis

The existence of super-regeneration has shewn us that regeneration is not merely the creating of substituent structures, but that its very nature involves the outward expression of a developmental potency still present in the adult individual. This appears still more convincingly in cases of heteromorphosis, in which at the same time the range of this potency is exhibited in a characteristic way. Heteromorphosis includes the appearance by "regeneration" of an organ of a different kind from the one that was lost, and the "regeneration" of an organ in a part of the body in which it is normally never found.

Heteromorphic formations frequently arise contrary to the polarity of the animal—understanding by polarity the existence of the two fundamentally different poles of the body. In sedentary animals, such as Hydroids, it is easy to distinguish these poles as apical and basal; in motile animals the oral pole, which generally is in front during locomotion, is opposite the aboral pole. In regeneration, polarity is as a rule strictly adhered to; that is to say, those organs are regenerated at a particular pole which normally belong to it. In certain circumstances, however, the regenerates may defy this polarity; for example, in a Planarian a head end may be formed instead of the amputated hinder end—a case of heteromorphic regeneration. Heteromorphoses shewing disregard of polarity occur in the most diverse groups of animals. But apart from the regeneration of organs of incompatible polarity, they may be manifested by the formation, in relation to subsidiary axes of the body, of organs which normally lie at one pole. Something of the sort is well shewn in the Anthozoa, where, if a wide-open wound be made in the cylindrical body wall, a new crown of tentacles grows round the wound.

The most interesting heteromorphoses, however, are those in which a substitute appears which is unlike what is lost, particularly in animals the early stages of whose germs have a strongly marked mosaic character—the whole of the germinal

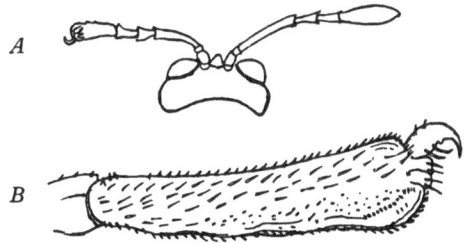

FIG. 68.—*Cimbex axillaris. A*, head with left antenna bearing a claw. *B*, another view of the terminal joint of the same antenna. (From Korschelt, after Przibram.)

regions in early development appearing to have exclusive potentialities. Very clear and beautiful cases are met with among Arthropods.

The oldest observation of this kind is one upon a Saw-fly, *Cimbex axillaris*, in which there occurred heteromorphic regeneration of the end of the antenna—which bore undoubted claws. That the origin of these claws is a regenerative process has been experimentally confirmed (Figs. 68, *A* and *B*). If, when the larva of this animal is about to pupate, its antenna is simply cauterized, the lesion being not too deep, a normal antenna is regenerated. If a greater lesion is produced, however, so that the effect of cauterization invades the part remaining besides the distal part, the clubbed end of the regenerating antenna forms an undoubted foot—with claws, pads, and the arrangement of hairs appropriate to a foot.

Heteromorphic regeneration of the antenna is possible also

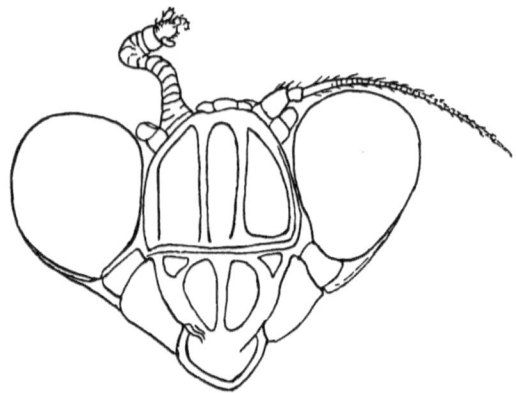

FIG. 69.—Head of *Sphodromantis bioculata*: the right antenna after having been cut off at the level of the scape has regenerated to form a heteromorphic fore-limb. (From Korschelt, after Przibram.)

in the Mantidæ. The antenna consists of two basal joints and a many-jointed flagellum. Amputation in the region of the flagellum leads to normal regeneration of the amputated part.

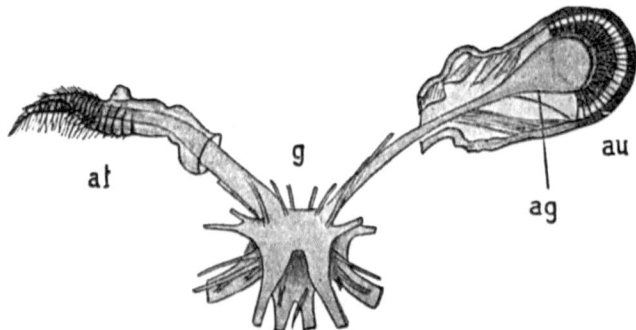

FIG. 70.—*Palinurus vulgaris*: regeneration of an antennula in place of an eye, after amputation of the eye-stalk together with the optic ganglion. *au*, eye; *at*, heteromorphic antennula; *g*, cerebral ganglion; *ag*, optic ganglion. (From Korschelt, after Herbst.)

In the case of amputation at the basal joint it is different: the regenerate is a foot which, in its characteristic bend, at least approximates to a prehensile limb (Fig. 69).

A beautiful case of heteromorphosis is provided by the decapod Crustacea. The eyes of these animals are set on short stalks. In the stalk there is found a part of the nervous system specially belonging to the eye, the optic ganglion (Fig. 70, *ag*). If the eye alone is removed, leaving the ganglion intact, it is regenerated, even though the normal condition is not always re-attained. If the whole eye-stalk, together with its contained ganglion, is amputated, an eye-stalk and eye are not regenerated, but an antennula (Fig. 70, *at*). Although this shews clearly that the nervous system affects the kind of organ regenerated, it is not known why the absence of a ganglion should lead to the formation of an antennula specifically. Similarly we do not know how it comes about in the cases already mentioned that a limb takes the place of an amputated antenna. We are principally concerned here, however, with the fact of heteromorphosis, and with its revelation of an unexpected range of potency.

In the occasional production by Arthropods of a structure typically belonging to another segment of the body, in place of an amputated appendage, we have what is called *homœosis*, so that in such cases we may speak of homœotic heteromorphosis.

III. A NON-PREFORMISTIC CHARACTERIZATION OF DEVELOPMENT

1. The Conceptions of Preformation and Evolution

The facts so far ascertained with regard to the behaviour of potency both in embryonic development and—after its completion—in regeneration, lead at once to an important though at first a negative conclusion regarding the nature of development. For it is a necessary consequence of the above class of facts—here illustrated by a few examples only—that development is not preformational in character.

The meaning of this will be best understood if we start historically with the origin and development of the idea of preformation. At one time the attempt was made to explain embryonic development by supposing that the various parts of the individual were preformed, as such, in the fertilized egg, and that development was nothing more than an increase in size, together with an unfolding, or evolution, of organs or parts of organs already fully formed in the earliest stage and, so to speak, "encapsuled." In this way development would proceed like the unfolding of a flower-bud, where all the parts of the flower are present, "fitted-in" to one another, but in an immature form. This early view of the preformation of the parts in the germ naturally could not survive direct microscopic investigation of the germ in its stages of development. In principle, however, the preformation of the organism in the fertilized egg was by no means abandoned on that account, but was altered so as apparently to accord with established facts.

The first modernization of the old preformation theory was Weismann's theory of the Germ-plasm. According to this theory every peculiarity of the adult organism is represented by a special corpuscular carrier of heredity, or *determinant*, which is present in the germ-plasm. It is present, that is to say, in the protoplasmic basis of development, contained in the germ-cells, later in the fertilized egg, and chiefly in the nuclei of cells. The primordial rudiment is thus not the organ itself; but there is for every organ and for each of its special parts a definite particle of protoplasm which represents a definite organ or part of an organ. This point of view has been adopted in essentials by modern genetics, at any rate in so far as that is represented by the extreme form of the chromosome theory. According to this there corresponds directly to each character, and each differentiation of the complete individual, a so-called *gene*, and this gene exists as a discrete particle of substance in a particular chromosome of the germ-cell nucleus. In reality a material gene of this sort is nothing more nor less than a corpuscular determinant in Weismann's sense. The whole of development is due exclusively to the action of the determinants—that is of the corpuscular genes, which are subdivisions of the chromosomes. According to Weismann the separate determinants are joined together into groups called *ids*. The earlier chromosome theory regarded the individual chromosomes as being aggregates, more or less closely bound, of a large number of corpuscular genes. In this way permanence, in the material sense, is ascribed to the aggregate of genes—the chromosome—during all the phases of cell-life.

Thus, all the structures which appear in development, are already actually and materially preformed in the germinal rudiment, though a complicated series of changes in the rudiments is necessary before the outward differentiation can be realized in its final form. In development, therefore, nothing new is formed: the whole diversity of the adult organism exists from the very beginning in the form of discrete parts of the protoplasm, and during development no new manifoldness arises. Each organ, each peculiarity of an organ, is directly preformed as a material particle situated in a particular part of the germ-plasm, or in more modern language, in a particular part of a particular chromosome. In this way development is purely and simply an unfolding of this complicated primary structure—a simple evolution. Whatever may be present in the fertilized egg beyond these preformed material rudiments is only a substratum, completely under the influence of the rudiments, and playing, therefore, a passive part at most. The point of view that has here been given in outline is not always perhaps so crudely expressed in modern literature, but such a clear-cut formulation is absolutely necessary if one is to understand the conception which actually lies at the root of all the less positive forms of the theory.

If the above be true, development must consist in the exact subdivision of the germ-plasm, and of its distribution to the individual cells of the embryo, and to the cell-complexes and the organs which later arise from them. The complication of primordia present in the germ-plasm of the egg is to some extent disentangled during cell-division—"developed"—each cell receiving in this way the special primordia or determinants for the cells descended from it. Development is an unfolding, and is initiated by the distribution of groups of determinants among the daughter-cells of the egg. This division is a qualitative one. The daughter-cells inherit from the egg unlike parts of the germ-plasm. The divisions of the egg and the succeeding cell-divisions are differential as regards heredity. Each cell thus carries within itself the conditions for its own further development; the development of the egg and its daughter-cells is a pure "self-differentiation." Development is a pure mosaic, since the end of development is determined by the beginning, each particle pursuing its course independent of the others. The cells of the "germ-track," giving rise as they do to the future reproductive cells, are not involved, however, in the differential distribution of the germ-plasm. These cells receive the germ-plasm intact from the egg, and the complete foundation for the next generation is thus conserved.

In outline these are the principles of development as Weismann, following out the conception of preformation, has stated them. The modern preformationist views, which are the foundation of the chromosome theory of heredity, do not generally speak of such a distribution of primordial units. If we insist, however, on the preformation of structures by means of discrete material particles, there is no alternative to the acceptance, with Weismann, of the view that the essential character of development is a differential distribution of these corpuscles to the various parts of the germ. To preserve those differences between individual germ-regions which are a primary necessity of every conception of preformation, it makes no difference whether the primordial units be sought in the nucleus only or also in the cytoplasm; neither is it of fundamental importance to assume that these material corpuscles, lodged in the nucleus, are distributed to all the cells with genetic uniformity, while the individual cells possess a fundamentally different cytoplasm. What is essential is that the individual germinal regions should not only show unlike potentiality, but should acquire unlike material primordia, whether by reason of differential nuclear division or of differential distribution of cytoplasm. Without some such difference in the primordial rudiments no theory of preformation is possible. On the assumption of a preformation of all the parts it is clear that each germinal region is strictly determined from the very beginning; any lack of certainty in its developmental fate is incompatible with this.

2. Inferences from the Facts of Experimental Embryology

Now the results of the experimental analysis of development shew that the potentiality of the blastomeres and of the parts of the germ is *not* wholly limited and exclusive, but that it is really greater than would appear from their normal fate in development. A number of examples in this connection have already been given, others will have to be mentioned later. The facts are that the $\frac{1}{2}$- and the $\frac{1}{4}$-blastomeres may possess the same potency as the whole egg, and further, that in certain circumstances a part of the germ may pursue a course entirely different from that which it takes in normal development. Developmental fate is thus not determined *ab initio* with absolute exclusiveness. This indeterminateness is present not only in early embryonic stages, but can easily be shewn to occur, in processes of regeneration, after the completion of the whole development. The potency of individual germinal regions is thus, to say the least, greater than their normal performance. It follows therefore that even after the many cell-generations intervening between the fertilized egg and the formation of embryonic and tissue cells, the mass of primordial rudiments received by each cell from its parent cell cannot have been so diminished by cell-division as only to suffice for the realization of a partially determined fate. On the contrary, this end attained in embryonic development depends, not upon a partial summation of the primordial rudiments—whatever they may be—but upon a complete summation of primordial factors and of those other factors, not immediately present in the egg, which appear in response to influences from within and without. Not only is the assumption of a qualitatively unlike cell-division inadequate to explain the phenomena in question: it directly contradicts the facts. It has been attempted to rehabilitate this qualitative or genetically differential division by assuming the presence —in addition to the real determinants or genes—of supplementary determinants, which determine the fate of the cells, and which are only activated in special circumstances; but this over-elaboration is in reality nothing more than a confession of the failure of genetically differential division to explain many experimental results. Less forced, and more in accord with the observed facts, is the assumption that division is genetically uniform, every cell receiving the full complement of factors, a part of which, by certain processes occurring in development, become determinative of the actual fate of the cell. In postulating such a qualitatively equal distribution, it is not necessary to assume that the tissue-cell in all circumstances possesses the full complement of rudiments. It would accord well with the facts to suppose that during specialization certain parts of the rudiment-plasm might here and there somehow disappear; this, however, not because an unequal distribution had taken place, but because particular complexes after genetically similar division had degenerated.

That development has not the character of preformation is already shewn by the phenomena of potency, but this is further confirmed by the frequency of our observation that developmental fate is indefinite. There cannot exist, then, for each individual part of the germ a predetermination, present once and for all from the beginning, in the preformistic sense. Not only the regulation of a partial embryo to form a whole, but metaplastic and heteromorphic regeneration also would be impossible.

For the moment we will content ourselves with this negative statement about the nature of development, reserving till later any positive characterization of the process.

3. The Same Conclusions as They Affect Mosaic Eggs

It might at first sight appear that the above conclusions could only be valid for regulation eggs; because in the case of mosaic eggs there is often a quite obvious division into qualitatively different blastomeres. But there is, in actual fact, no such limitation of the non-preformational character to one particular group of eggs.

We are dealing here with a fundamental property of the process of development; it must be assumed not that there is an antithesis between two fundamentally different types of development, but that the process of development has essentially the same character in all animals. The power of regulation is by no means entirely lacking in mosaic eggs, though it may appear in them in widely varying degrees. A capacity for regulation, however, always denotes a range of potency going beyond the normal fate, and therefore, also, a greater content of primordial rudiments. The appreciable power of regeneration possessed by some forms which are developed from mosaic eggs points in the same direction. Further, every possible gradation exists between typical regulative eggs and extreme mosaic eggs; there is no sharp dividing-line between the two types, though the extremes can easily be distinguished. Regulative and mosaic eggs can in fact be arranged to form a continuous series, in which the special cases are joined by intermediates. Such a series beginning with typical regulative eggs and ending with extreme mosaic eggs would be roughly set out in this order: Amphioxus, Teleosteans, Mammals, Nemerteans, Urodeles, Anurans, Ctenophores, Annelids, Molluscs, Arthropods, Nematodes, Ascidians. The ideal extremes—the purely regulative egg and the purely mosaic egg—are nowhere realized in nature. Certain peculiarities of development in typical regulative eggs have a mosaic character; and, conversely, regulative characteristics are not entirely lacking in mosaic eggs. Which of the two types of development is emphasized depends simply on the particular species under consideration.

If in the cleavage of the typical mosaic egg differences between the germinal regions appear unusually early, that is not because the laws which govern them are different from those governing regulative eggs, but because the time at which determination occurs is different in the two types of eggs. In regulative eggs the fate of individual germinal regions is not decided until comparatively late, often not until the germ is well advanced in cleavage. Therefore the individual cells are not differentiated until rather late, and then only gradually. In the mosaic egg it is otherwise. The processes of determination are completed comparatively early, in certain cases even before the beginning of cleavage, and as a result the effects of determination quickly appear, so that before or at least during cleavage the separate protoplasmic regions are differentiated. Upon the time when determination occurs—and therefore when differentiation begins—depends that diminution of potency which appears in mosaic eggs during cleavage and in regulative eggs somewhat later. The distinction between the two thus consists in a chronological displacement of the different processes of development relative to one another, since diminution of potency is not the same as diminution of primordial rudiments. It need hardly be said that power of regulation and uncertainty of fate decrease hand in hand with the progress of determination. The qualitative subdivision of a mosaic egg is not a primary *modus operandi* of differentiation, but merely the secondary consequence of a differentiation present already as the result of precocious determination. If there existed in the germ-cell a pure mechanism of preformation, a qualitatively different subdivision would be necessary as a means of differentiation; but actually there is no such mechanism. But though the distinction between regulative and mosaic eggs is only one of degree, it is nevertheless inadvisable on practical grounds to discontinue the use of these terms.

Something remains to be said about the regulation of a partial germ to form a whole animal. It might be imagined that this regulation was brought about by the ½-blastomere assuming at the outset the structure of the whole egg—its proportions alone being reduced. This, however, is not the case. The explanation is rather that the separated ½-blastomeres still consist of similar protoplasm, the fate of whose individual regions is not yet strictly determined. This is approximately true of regulative eggs, though never does it take this ideal form. Rearrangement of materials in the isolated cells on the plan of the whole egg is in this way unnecessary; the structure of the cytoplasm can indeed be greatly disturbed without impairing its capacity to produce a whole. In the mosaic egg, on account of the early determination of individual regions, a qualitative difference between the blastomeres is generally present; their special role is already marked out. A blastomere may lack a specific part of the cytoplasm necessary for the formation of a whole embryo, as, for instance, in the egg of *Dentalium*, where the polar lobe is allotted to one particular blastomere, and where, if a part of this pole plasm goes into another blastomere, this latter can perform just as much as the normal cell from which the pole plasm has been taken. For the production of a whole out of a part it is only necessary that there should be represented in the part all the determined —we might add all the determining—regions of the cytoplasm; or, conversely, that no steps should have been taken towards determining the individual plasma-regions, as in the case of the typical regulative egg. Whether the arrangement of material corresponds to that of the unsegmented egg is of secondary importance.

CHAPTER V

THE PROBLEM OF DETERMINATION

I. The Chronology of the Process of Determination

1. General Significance of Determination

When the developmental fate of a part of a germ is decided—
in other words, when its *determination* is achieved—a limitation
of its potency is involved as a matter of course; for determination always means that the course of subsequent development
is more or less rigidly laid down. Therefore range of potency
furnishes a criterion of the condition existing as regards
determination. The greater the potency, the less marked will
be the degree of determination, and the converse is true; so
that all that has been said in the previous chapter about the
potency problem applies with equal force to the determination
problem, if the same point of view be adopted.

The completion of determination is not always easily
recognizable by definite, visible changes; but once a part of
the germ that has been determined begins to follow its own
definite course a special differentiation quickly sets in. The
beginning of differentiation and the completion of determination are so near together that in practice the two processes can
hardly be separated. Thus, a germ whose parts are already
strictly determined will generally also shew appreciable differences in the differentiation of its individual regions. To this
extent the state of differentiation of the germ is an index of the
degree of its determination.

Determination of all of the germinal regions never takes
place at one definite moment; its establishment is spread over a
considerable space of time. The length of time required cannot
be stated in absolute terms, but only relatively; and the period
during which determination is taking place varies very much
in extent in different groups of animals. Gradual progress from

PLATE IV

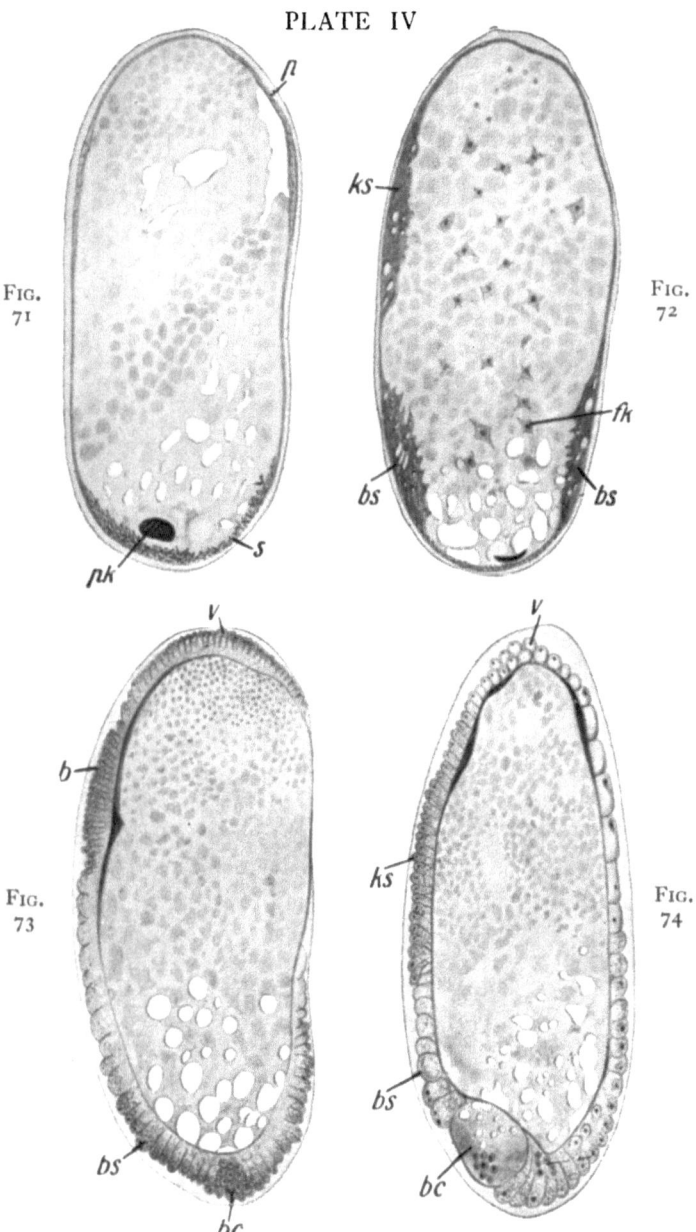

FIGS. 71 to 74.—*Camponotus ligniperda*. Unsegmented egg and young embryos in longitudinal section. Fig. 71, egg just laid, rudiment of blastoderm uniform throughout its whole extent. Fig. 72, formation of regions in the cytoplasm of the blastoderm-rudiment; inner cytoplasm with many cleavage nuclei. Fig. 73, blastoderm composed of cells and plainly shewing different regions. Fig. 74, later stage, in which the regions are still more definite. (After Hecht.) *P*, superficial layer of cytoplasm (blastoderm-rudiment); *pk*, polar mass of cytoplasm (*Polkörper*); *s*, symbiotic bacteria; *fk*, cleavage nuclei; *ks*, germ-band; *bs*, blastoderm syncytium; *b*, blastoderm; *v*, cell region at anterior end of germ; *bc*, bacteriocytes.

the general to the particular is characteristic of the determination process, and therefore of development as a whole. At the beginning of the process determination is only general, so that the whole is defined from the outset while the individual parts are only secondarily established. The whole organism is therefore not the sum of a number of parts originally present: it is the whole which is primary—the parts are secondary subdivisions. This of course is simply an affirmation of the non-preformational character of development.

Development is carried out by a special *reaction-basis*—if we give this convenient name to the whole of the specific constitution of the germ-cell, including those of its primary morphological differentiations which are essential to its development. The reaction-basis is the actual bearer of all the internal factors of development which are inherited. It has already been pointed out, however, that it does not contain a preforming particle of protoplasm corresponding with each single part of the resulting individual. It is more true to say that the egg—i.e. the individual in its unicellular condition—owes its general determination as regards purely specific and racial peculiarities to the reaction-basis, while determination of the morphological value of the separate parts follows step by step. The solution of the problem of the specific and racial determination of the whole organism would therefore involve an investigation of the real nature of the reaction-basis. Experimental embryology, as an analytical science, cannot, however, argue from the whole to its parts, but must arrive at a comprehension of the whole through understanding the parts. Hence, from the point of view of analytical embryology, the question of the determination of the parts is the essence of the determination problem.

This determination comes about by degrees, earlier or later according to the kind of animal in question, so that an indeterminate condition gradually gives place to a determinate one. Further, since all the parts are not determined at the same time, there arise in the germ local and spatial differences in the state of determination. But determination in itself is a gradual process. Not only does it lead little by little from the general to the particular, but the rigidly determined condition is connected by intermediate stages with the indeterminate condition, beginning with the first appearance of determination —when it is still indefinite—and proceeding to its final complete establishment. In order to distinguish between these degrees of determination we may use for the introductory phase the term *institution*, and for the final phase—in which ontogeny no longer remains uncertain—the term *destination*.

2. Determination and Cleavage

That the process of determination is essentially a gradual one appears especially clearly when we inquire at what moment in a particular ontogeny the character of the individual parts is settled. Now, it is obvious at the outset that the beginning of determination by no means coincides with cleavage. On the contrary, the processes are so independent of one another that it may safely be said that the division of the egg into blastomeres is not an agent of determination. To avoid undue repetition, we will simply recall what has already been said about potency in extreme mosaic eggs. Here the potency of individual regions becomes limited extraordinarily early, often even before cleavage, which shews that determination has already begun. This is demonstrated too by the appearance of visible differentiation of the various regions of the cytoplasm. At the same time, as we have said, there still remains in these eggs a certain degree of freedom for the action of regulative processes, so that even in these extreme cases determination does not end abruptly, but is completed gradually, sometimes not until after the close of cleavage. For the rest, there may be in all mosaic eggs a stage in which the determination of the parts has not yet taken place, though this stage (as against what happens in regulative eggs) may perhaps occur very early, even during pre-development.

Leading up to those germs in which determination begins during or after cleavage, there are interesting transitional cases. In these, when cleavage has begun by the division of the zygote nucleus, the cytoplasm has not yet undergone division into regions corresponding to the individual cells. It has nevertheless developed very clearly marked differences of character in its various regions, indicating the fact that the several fates of these regions are already marked out. As an example of such a case we may take the egg of the ant, *Camponotus ligniperda* (Pl. IV, Figs. 71–74). Immediately after laying (Fig. 71), the superficial stratum of cytoplasm, or blastoderm-rudiment (*p*), is equally thick and equally differentiated all over; at the hinder end lies a cytoplasmic mass (the *Polkörper*), which is perhaps a pole plasm (*pk*); and there are also present symbiotic bacteria (*s*) which need not concern us here. When later, however, the egg nucleus divides, and by repeated divisions forms the so-called cleavage nuclei (*fk*), another picture presents itself (Fig. 72). The surface layer of the egg now plainly exhibits different regions marked by the unequal thickness of its cytoplasm. Later stages prove that these regions possess different—and in most cases quite definite—fates in development. The nuclei at this stage do not yet lie in the cytoplasm of the blastoderm-rudiment; they are indeed far removed from it. One of the regions mentioned (*ks*) represents the first rudiment of the germ-band, the most important part of the embryonic rudiment. The other, forming a girdle round the hinder end of the egg, is the basis of the blastoderm syncytium (*bs*). Once the nuclei have invaded the superficial cytoplasm, thereby forming the actual blastoderm (Fig. 73, *b*), the regional differences previously established become easily recognizable. At the anterior pole of the egg lie cells rich in yolk-granules and in vacuoles, and which soon lose their epithelial character (Fig. 73 and 74, *v*). Some of these go to form an incomplete embryonic membrane, some later degenerate. The cells of the presumptive germ-band (*ks*) are more or less uniformly columnar, while those which pass into the blastoderm syncytium (*bs*) fuse later to form multinucleate giant cells. At the hinder end of the germ there are still found bacteriocytes (Fig. 73, *bc*), which absorb the symbiotic bacteria and, later, degenerate after giving up their symbionts.

In a word, there is in this case a very early formation of presumptive organ-regions, the progressive determination of which arises by the precocious appearance of visible differentiations of the blastoderm-rudiment. This does not mean that in the egg which has just been laid, and which still possesses a homogeneous surface layer, no determination of special regions has yet occurred or is at least begun; it is very probable on the contrary that the first phase of determination has indeed begun, and that the visible differences later produced in the egg regions are a consequence of this. In any case, however, distinct organ-regions are already defined at a time when the cytoplasm still shows no sign of cleavage, though nuclear division has now produced a large number of blastoderm nuclei which later invade the different zones of cytoplasm. The fact that determination is independent of cell-division is here beautifully illustrated.

The gradual march of determination is again specially clear in eggs in which at least the main part of the process does not take place until after cleavage; these conditions are generally found in regulative eggs. There is perhaps no case in which some progress in determination has not been made before cleavage. For example, even in the sea-urchin egg itself, where the $\frac{1}{2}$-blastomeres possess so great a capacity for regulation, there is at the beginning of cleavage a certain difference between the animal and vegetative halves of the egg as regards their potency, or in other words, the degree of their determination. When, then, it is said that determination only occurs after cleavage, what must be understood by this is that it sets in principally during and after cleavage. In such cases the process of determination is protracted and its gradual progress is the easier to follow.

Each $\frac{1}{2}$-blastomere of the sea-urchin egg can itself produce a whole embryo; it cannot therefore be regarded as finally determined, since its function in development is still not defined. The same can be said of the $\frac{1}{4}$-blastomere, for it again produces a complete pluteus. The $\frac{1}{8}$-blastomere, however, is different, for if one of these be isolated, only a gastrula arises from it, and this cannot develop further. The isolated $\frac{1}{16}$-blasto-mere can only produce an incomplete gastrula; and, finally, blastomeres of the 32-cell stage, only a blastula. Though these cells still show no pronounced cytological differences, their condition with regard to determination is different, as is obvious from their different potentialities; in fact, while their potency gradually decreases, the degree of their determination increases. That determination is still not complete in the 32-cell stage is shown by the fact that an isolated blastomere of this stage develops no special organs, or parts of organs, but produces only a blastula which does not develop further. Here again it is seen that determination at first is general in character —in fact, that in the first stages of determination we are not concerned with the definitive and specialized laying-down of organs or even of rudiments of organs, but rather with the general planning-out of the presumptive organ-regions, within which regions a specialized determination takes place later. Similar observations have been made in the case of other regulative germs by means of isolation experiments.

3. Determination in the Egg after Cleavage

Further light is thrown on the progress of determination by investigating the potency-relations in early stages of the germ after cleavage. This can be done by subjecting the parts of the blastula and the gastrula to appropriate tests.

The fact that determination in the blastula of the sea-urchin is at first only of a generalized character is shewn by the capability of two normal blastulæ to unite to form a single giant blastula. This giant blastula produces a giant larva, whose organs are present in the same number and proportion as in normal larvæ. Since in this process many cells must pursue a course different from that which is normal to them, no detailed determination of organs can exist in the young blastula.

There is much evidence with regard to the potency (and therefore the degree of determination) of the parts of the blastula and gastrula in Amphibia, especially in the case of newts. Researches in this field have also shewn very clearly the existence of the indefinite phase and the definitive, fixed phase of determination—the phases, as we have called them, of institution and destination.

In order to test the degree of determination and the potency of various regions of the blastula and gastrula of *Triton*, transplantation experiments are employed, in which certain parts of the germ are exchanged. It is best to arrange these transplantations so that the donor and the recipient of the transplant differ in colour, since this facilitates subsequent observation. It is rendered possible by the dissimilar pigmentation of the germs of different species of newts, some eggs being almost free from pigment, while others are strongly coloured.

Figs. 75, *a–d*, Plate V, shew the kind of transplantation referred to. A circular piece taken from the roof of an almost pigment-free blastula of *Triton cristatus* is inserted just above the cranial (dorsal) blastopore lip of a strongly pigmented young gastrula of *Triton alpestris* (Fig. 75, *a*). The first sign of the blastopore can be seen (low down in the figure) as a dark groove. Gastrulation proceeds at first normally, until the transplant reaches the edge of the blastopore-lip; it is then, however, obviously arrested for a time in the median region of the dorsal lip (Fig. 75, *b*), while it proceeds without hindrance laterally. As a result the blastopore rim, which at this stage is normally crescentic, is drawn out to a point in the middle. Later the germ overcomes this hindrance (Fig. 75, *c*), and now the transplant itself is rolled into the interior of the gastrula, and finally disappears altogether from the surface (Fig. 75, *d*). In this way it passes into the median part of the roof of the archenteron. Operations of this sort can naturally be varied in many ways according to the end in view.

In *Triton* it is possible, even in the blastula, to distinguish three regions according to the character of the yolk-granules they contain—the animal field, the marginal zone, and the vegetative field—corresponding to the previously mentioned presumptive regions of the germ-layers. These already possess a certain degree of determination, though their fate still shews so much uncertainty as to demonstrate the incompleteness of this. Now if a small piece of presumptive ectoderm from a young gastrula is implanted into the vegetative field of another gastrula so that it lies in the region of the future yolk-plug, we find that it is not incorporated into the endodermal material but forms a separate epithelial mass. This resistance to incorporation may be due to the tendency of presumptive ectoderm to increase its surface. Material from the marginal zone, implanted in the presumptive ectoderm, sinks down and by its extension pushes out the surface. Parts of the vegetative field when transplanted into the animal field behave somewhat passively and are generally overgrown. These different tendencies of the presumptive germ-layer regions obviously play an important part in normal gastrulation; they show that particular regions are already to some extent specialized and have attained a certain degree of determination. But though germinal regions already influenced by processes of institution normally give rise to definite systems of organs, their fate may still be uncertain; in other words, determination in this stage is not yet in its final phase. For example, if a small, darkly pigmented piece of the presumptive medullary material of *Triton tæniatus* is implanted into the region of presumptive epidermis of a young gastrula of *Triton cristatus* (Fig. 76), it is easy to see that the implant, instead of undergoing its normal development, becomes itself part of the epidermis (Fig. 77). Conversely, in the reciprocal exchange, presumptive epidermis can be caused to form medullary-plate material.

A few further experiments of this nature may be mentioned. A piece of presumptive ectoderm from an early gastrula of *Triton*, implanted into the vegetative field of another gastrula, in such a way as to be carried into the interior during gastrulation, may give rise, according to its final position, either to the roof or to the side wall of the gut. If it is implanted in the area of presumptive mesoderm, it may form mesodermal products such as somites and lateral-plate mesoderm (Fig. 78). Presumptive mesoderm derived from the caudal (ventral) lip of the blastopore, where the tendency to invagination is least marked (so that morphogenesis is little affected by its removal), may in certain circumstances and in an ectodermal region develop into ectoderm.

Determination makes further progress during the gastrula

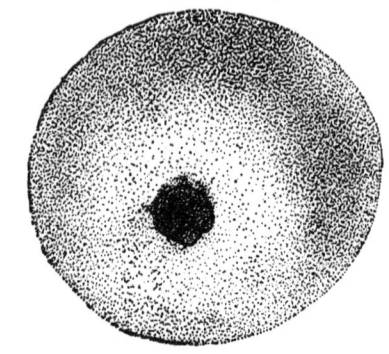

FIG. 76

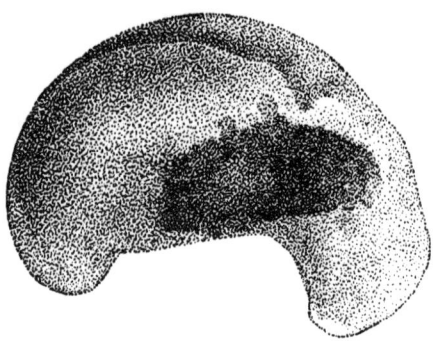

FIG. 77

FIGS. 76 and 77.—Germ of *Triton cristatus* (pale) with an implant of presumptive medullary material from *T. tæniatus* (dark), shewn at two different stages. Fig. 76, young gastrula of *T. cristatus*; implant of presumptive medullary material of *T. tæniatus* in the region of presumptive epidermis—shortly after the operation. Fig. 77, the same embryo seen from the side at a later stage; medullary plate above in the figure; the dark implant has become epidermis. (After Spemann.)

stage. At the end of the process of gastrulation presumptive epidermis and medullary material are no longer interchangeable; for if pieces of material are exchanged, as in the above-mentioned experiments, they no longer develop in accordance

PLATE V

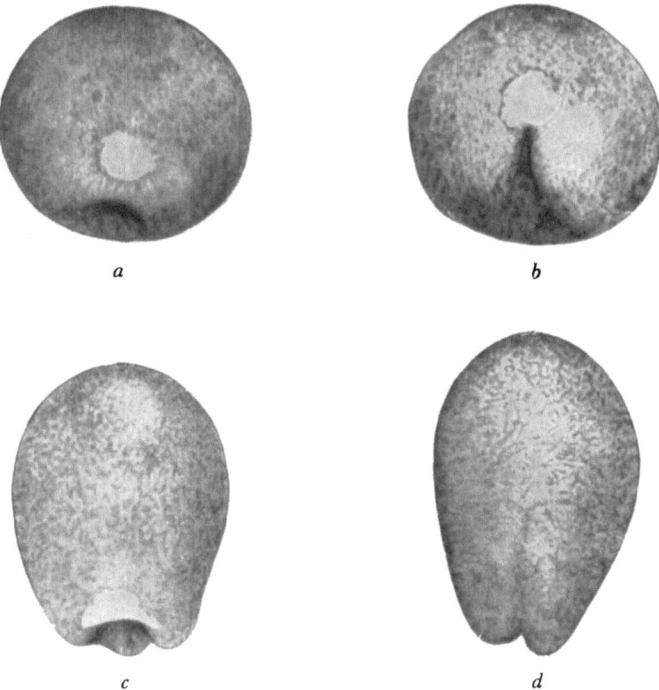

a *b*

c *d*

FIG. 75, *a–d.*—Behaviour of a piece of a *Triton cristatus* blastula (pale in colour) transplanted into a gastrula of *T. alpestris*; rolling-in of the transplant during gastrulation—the same germ shewn in different stages. Age of these stages counting from the time of the operation: *a*, 50 minutes; *b*, 6 hours; *c*, 23 hours; *d*, 2 days. (After Spemann and Geinitz.)

with their position, but in accordance with their origin; from this time on they are finally determined. Pieces of the presumptive medullary plate, taken from a gastrula with a yolk-plug or with a slit-like blastopore, still develop, in the region of the epidermis of another germ, like their new surroundings. Corresponding transplants, from a stage with closed medullary folds are, however, no longer indeterminate, and always develop according to their place of origin. In the older gastrula,

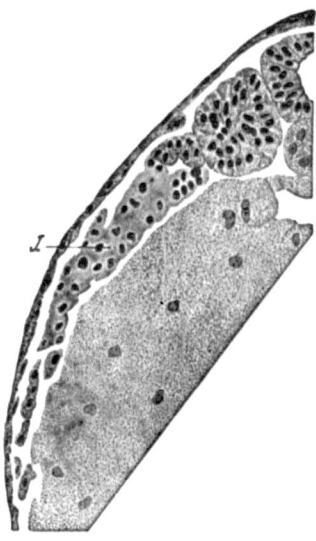

FIG. 78.—Part of a transverse section through an older embryo of *Triton alpestris*. The lateral mesoderm (*J*) is derived from presumptive ectoderm of *T. cristatus*. (After Mangold.)

determination has already become very specialized—which is not the case in the somewhat younger germ. This specialized determination already discriminates particular organ rudiments. Implantation experiments of this kind shew that in the more advanced gastrula of *Bombinator*, the liver, the pancreas, and the gut are already determined in detail, and their relative positions fixed.

The same phenomenon—determination in its main outlines, followed by a specializing and final fixing of its character

—is reproduced on a smaller scale, but in a substantially similar way, within the separate organ-rudiments in later embryonic stages. In the case of organ-rudiments also, there is at first a period during which determination has set in but is not definitive: during this time changes are still possible.

In the Anura, the young, just-projecting posterior limb-bud is determined in a general way as a limb rudiment, but within this rudiment developmental fate is not defined in detail. Thus, half of such a bud may produce a whole limb. An alteration in its symmetry-relations is also possible, so that a left limb can arise from a bud of the right side after a suitable transplantation. What comes first is the determination of the axes of the cross-section; next, determination in the direction of the long axis. The polarity given to the rudiment by the latter, however, is reversible even when the determination in cross-section is completely fixed. Without citing more individual cases, we may say that a similar behaviour is found in the rudiments of other organs, and it is only necessary to repeat that the time of final determination varies in different organ-rudiments.

Regeneration blastemas, which often must be regarded as analogous to rudiments of organs, are also determined, at first, only in a general manner, the special determination of their parts appearing only gradually.

Thus, to speak generally, the condition of a germ as regards determination shews chronological differences. There are also, as we have pointed out, spatial, or local, differences. These local differences manifest themselves in experiments in the form of differences in the potency of the various parts of the germ. For example, the animal half of the sea-urchin blastula cannot produce a gastrula, but the vegetative half can; thus the two halves must be different, and the existence of this difference is due to the fact that determination has not progressed equally throughout the blastula. If a young germ of *Triton* is divided into a dorsal and a ventral half, a fairly normal embryo is obtained from the former, but only a defective structure from the latter—a further proof of the existence of local differences in determination.

Thus it happens that in the same germ there may exist, side by side, parts already fully determined, and others in which determination is still indefinite, possibly not even begun. In this connexion we need only compare the limb-buds of the Amphibia with the condition of other regions of the larva. Since differentiation is directly connected with determination, local differences in determination are manifested by an unequal degree of differentiation in the various regions of the germ. This is very clear in the early stages of mosaic eggs; in regulative eggs it is generally not recognizable until rather later. It is very clearly demonstrated by such results that the process of determination is a gradual progress. It is seen that the egg does not contain a ready-made diagram, so to speak, of the individual which is to be formed, but that this diagram itself is only developed little by little—a process of an unmistakably non-preformational character.

II. SUBORDINATE ORGANIZATIONS OF THE EARLY EMBRYO AND THEIR RELATION TO DETERMINATION

1. Polarity of the Egg

The real achievement of development is not the formation of the organs in detail, but the creation of an individual; that is to say, not the production of parts but the differentiation of a whole. Specification of the developmental fate of a part always takes place within the framework of the whole and with reference to the whole. Determination is, in fact, a function of the whole germ, and the essence of development consists not in piecemeal action but in integral action. This is true not only of ontogeny proper, but of the process of regeneration. There are, indeed, cases of regeneration in which the unity and integrity of the individual are to a certain extent broken down (heteromorphoses and the like); but in many other cases the relation of the whole to the determination of the regenerate is undoubtedly of capital importance. The component parts of the whole germ play an important part, however, in the realization of development. Since the parts ultimately present are not preformed *ab initio*, and since their determination advances by insensible gradations, the first products of development are not definitive organ-rudiments, but provisional organizations of the various parts of the germ, which only secondarily acquire their functions of determination. The farther development progresses, the more do the subdivisions of the germ become responsible for special duties, but always within the framework of the whole and without breaking away from the integral system. Since these subdivisions begin to influence the other regions of the germ, and so to affect the process of determination, these subsidiary organizations themselves become organizers. The importance of the parts at a later stage of development becomes diminished.

Among the earliest organizations of the egg is the production of its polarity in relation to the primary axis passing from the animal to the vegetative pole. That is not to say that this is always manifested in the form of visible differentiations, although it is often recognizable as a special arrangement of substances with reference to the axis. This is seen at its simplest in telolecithal eggs. In general, polarity exhibits itself in the fact, among others, that the products of the maturation-divisions are separated off at a particular point on the egg—so that they can with justice be called polar bodies.

The polarity of the egg generally originates extraordinarily early. In the egg of *Ascaris*, for example, it appears during the growth period of the oocyte, and obviously in connexion with the unilateral attachment of the egg to its nutritive apparatus, the rachis. At first no differences in the substances at the two poles can be demonstrated, though these may later appear; there must be assumed to be originally present a gradient of physiological condition from animal to vegetative pole. In the egg of the sea-urchin, *Paracentrotus*, polarity can, again, be traced back to the oocyte. Conditions in the egg of *Polyphemus* are similar; here also polarity is related to the topography of the original arrangements for nutrition. The behaviour in this respect of eggs like those of mammals, which mature in enclosed follicles of uniform thickness, is at present imperfectly known. Though polarity is established at different times in different kinds of eggs, there is nevertheless in every case a consideration connected with the fact of this polarity which affects the determination of the regions of the egg. For, on the one hand, the direction of cleavage (and therefore the subdivision of the egg) is related to the direction of the primary axis; and on the other hand, the primary polarity—perhaps at first merely physiological—leads to a visible polarity of differentiation, and to consequent differences in the character of the blastomeres. Further, it appears that the polar differentiation of the egg results in unequal powers of regulation in the different regions of the egg, and of the blastomeres which they severally produce. This, however, means simply that there is a limitation of potency, correlated with a particular condition of determination.

In another aspect also polarity appears of great importance. In development we are not concerned merely with the fact that certain differentiations are formed, but that they arise in a particular place. This localization problem is at least as important as the determination problem. Polarity, though it may be far from obvious, is a first step towards a definite topography of the differentiations of the egg. Particular organ-rudiments are, of course, not formed by the establishment of polarity. Polarity is a matter, rather, of the creation of the first subsidiary organization of the germ, the influence of which on the localization of organ-rudiments later to appear is but an indirect one. The degree of polar anisotropy of the cytoplasm of the egg varies greatly, as does also its effect on later development—this effect runs roughly parallel to the series of intermediates between typical mosaic and regulative eggs.

2. Pole Plasms

True polar differentiation is shown especially clearly in eggs, such as those of Molluscs (e.g. *Dentalium*) and of many worms (e.g. *Tubifex*), which form so-called pole plasms. In the Oligochæt, *Tubifex rivulorum*, these consist of collections of cytoplasm with a definite configuration at the two poles of the egg (Fig. 79). During cleavage, which is of the complicated spiral type, they pass completely into one of the ¼-blastomeres, the *D*-cell, from the descendants of which arise the somatoblasts, and subsequently the ectodermal and mesodermal germbands. In the egg of *Dentalium* the pole plasm of the vegetative pole produces the "yolk-lobe," which finally unites with one of the ¼-blastomeres, as we have already seen (Cf. p. 94).

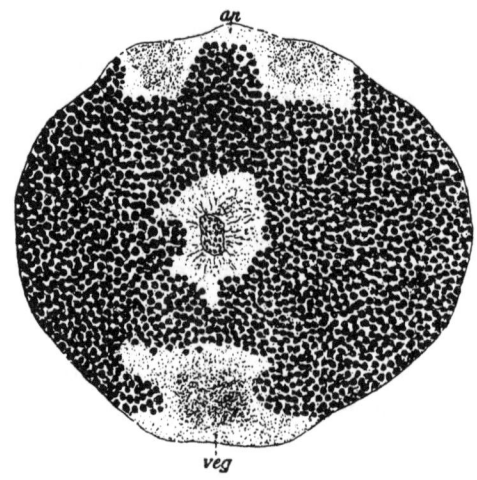

Fig. 79.—Ripe egg of *Tubifex rivulorum* shewing pole plasms at the animal pole (*an*) and the vegetative pole (*veg*). (After Penners.)

The importance of pole plasms in connexion with the limitation of the potency of blastomeres has already been mentioned. Only that blastomere to which the pole plasm is allotted is able to regulate itself and form a whole embryo. The allocation of the pole plasm thus marks a stage in the progress of determination, since the limits of potency of the other blastomeres are thereby narrowed down. Separation of the pole plasms and their union with a particular cell does not, however, give a completely specialized determination. For the *D*-cell while still forming part of the whole germ produces less than when it is isolated, since in this latter case it can produce a whole organism. This means that the relations between the blastomeres are important in the directing of subsequent development, and further that the appearance of pole plasms does not involve the complete preformation of organ-rudiments. Indeed, the presence of the whole of the pole plasm is not essential for the formation of a complete individual from part of a germ; about half of the pole plasm suffices for this. We must conclude, then, that pole plasms do not constitute a direct mechanism of preformation.

The presence of these pole plasms is a necessary condition for the appearance of particular parts of the germ and also of particular organs. They have therefore been called *organ-forming substances*. But they are not "organ-forming" in the sense that the tissues and organs in question are formed directly from them, or that they directly preform these organs: we are not dealing with actual organ-rudiments. Their relation to the final differentiations is, on the contrary, only an indirect one. This is proved by the fact that reduction in their quantity, and their diversion into other blastomeres, is possible without necessarily preventing normal development. The importance of the pole plasms is, rather, that they influence in a definite way the direction of development of the cells, and thus determine their fate. In normal development they determine cell *D* and its descendants, but if they are made to pass into another cell, they exert the same determining action on this. In any case, one thing is clear: that with the appearance of differences in the germ (in this case the unsegmented egg) special subordinate organizations begin to play a definite part in determination, since it is in them that the decision of developmental fate originates.

The formation of special regions of the cytoplasm is naturally most strongly marked in the mosaic egg. But since no strict dividing-line can be drawn between regulative and mosaic eggs, it would be expected that even in the former there would be a separation between special regions acting as bearers of determination. This is actually the case: in eggs of a strongly regulative character the separation of special cytoplasmic regions can be seen to occur. These assume a somewhat similar role to that of the pole plasms of the mosaic eggs we have mentioned, that is, they are the source of processes of determination. The evidence is particularly abundant in the case of the Amphibian egg, which, though it is not a regulative egg in the strict sense, conforms much more to this type than to the mosaic type.

In the ripe Amphibian egg polarity is obvious and is manifested by the direction of the first cleavage, in the plane of which normally lies the primary egg-axis. For the rest, no special protoplasmic differences are recognizable at first in the cytoplasm, apart from the disposition of its inclusions. In the egg of *Rana fusca* (*R. temporaria*) this disposition is altered to some extent by fertilization. At first it is only possible to distinguish a darkly pigmented animal pole and a pigment-free vegetative pole. About two hours after insemination, however, a change appears in the distribution of pigment, not very striking but nevertheless quite distinct, in that a third zone can now be recognized on the surface of the egg—the so-called grey crescent (Pl. VI, Fig. 80, *a–d*). Before its appearance, the distribution of pigment is uniform all round the egg, but the pigment from now on partly disappears from a crescentic area below the equator, just where the white vegetative field passes into the pigmented region of the egg, so that a lighter field arises, which embraces about half the circumference of the egg as a crescent. Though it is not always equally distinct, in favourable cases this can still be recognized in the blastula. The grey crescent is the expression of the bilaterally symmetrical structure henceforward possessed by the egg; the plane which includes the primary axis of the egg and the middle of the grey crescent is the plane of symmetry of the embryo. The middle region of the grey crescent is the place where, later on, the dorsal, or anterior lip of the blastopore lies, and this last determines the position of the axial organs (notochord, neural tube, somites). Thus in the vegetative half of the frog's egg a special plasma region is differentiated, and, as in the case of

PLATE VI

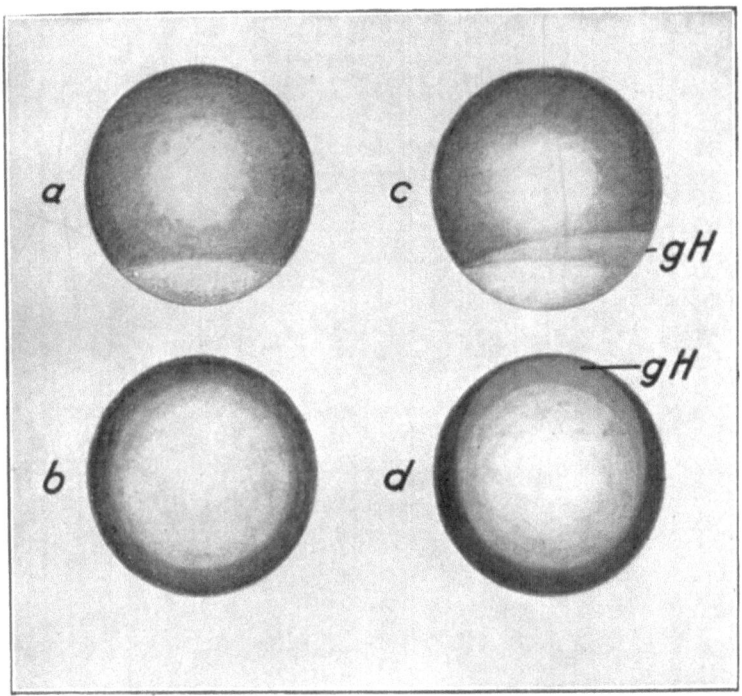

FIG. 80, *a–d.*—Diagrams shewing the grey crescent in the egg of *Rana
fusca a*, egg seen from the side, and *b* from the vegetative pole before the
formation of the crescent; *c* and *d* corresponding views after its formation
—*gH*, grey crescent.

the pole plasms mentioned above, with this region special processes of determination are later connected.

3. Organizers in the Egg after Cleavage

The substance of the grey crescent is the material basis of those blastula cells which later arrange themselves to form the first rudiment of the blastopore lip. It can therefore be called the presumptive region for the dorsal lip. As a result of this relation it is possible by investigating the part played in development by the anterior lip of the blastopore, to discover the importance of the cytoplasmic differentiation of which the grey crescent is the visible sign.

The anterior lip of the blastopore, at first a subordinate organization of the germ, now itself becomes an important *organizer*; that is, there pass out from it influences which organize other regions of the germ. These influences play an important part in specifying the development of the regions in question and are thus involved in determination.

The formation of the anterior lip of the blastopore take place as follows: an indentation of the wall of the blastula begins to carry the presumptive chorda-mesoderm material under the presumptive medullary region. This process is continued and completed during gastrulation by the invagination of chorda-mesoderm material. The underlying chorda-mesoderm material exerts from the outset, however, a definite influence upon the germinal material above it, so as to *induce* in this the formation of a neural plate and with it the other organ-rudiments which lie in the axial region of the embryo. In the lip of the blastopore, and in particular in the cellular material which is there invaginated, we have, therefore, a source of determination exercising a definite influence on the developmental fate of other germinal regions.

That the lip of the blastopore plays the part of an organizer or as it has been well called "organization-centre" can best be proved by transplantation experiments. In these a piece of the organizer is inserted into a region which has no organizer—

either in the same or another germ. The transplant now exerts the same action as it does when developing normally in its normal position.

An experiment of this kind is illustrated in Figs. 81–84. In this case a piece of the organizer just over the blastopore —that is to say a small piece of chorda-mesoderm material not yet invaginated—was taken from a gastrula of *Triton cristatus* and implanted into the left side of a *tæniatus* embryo of the same age. In later development nothing more is seen of this implant; in the position it occupied there appear two elongated ridges, surrounding a slit-like depression. These ridges gradually increase somewhat in length, and draw close together superficially (Figs. 81 and 82). We have here in fact secondary medullary folds, which have been induced by the organizer. The primary medullary folds have completely developed in the meantime along the whole length of the embryo, and begin to close dorsally (Fig. 81). While the embryo itself develops further and forms its optic vesicles, auditory pits, somites, and tail-bud, the secondary medullary folds close to form a medullary tube (Fig. 83). But the induction of such a secondary neural rudiment is not all: near it secondary somites are laid down; at the anterior end an auditory pit appears on either side, at the same level as those of the definitive embryo (these are shewn in the figure by small round spots); further, the hinder end of the secondary embryo is lifted up clear of its host to form a projecting tail, and a pronephric duct can be made out. Thus, the action of the implanted fragment of organizer has been to induce the formation of a complete secondary embryonic rudiment.

All this is fully confirmed by the study of series of transverse sections (Fig. 84). The primary embryonic rudiment shews the axial organs in their normal morphological relations (medullary tube, notochord, somites, pronephros, optic cups, and vesicular auditory rudiment with a suggestion of the ductus endolymphaticus). In the secondary embryonic rudiment the axial complex of organs is well established throughout, though not so fully developed as in the primary embryo. The

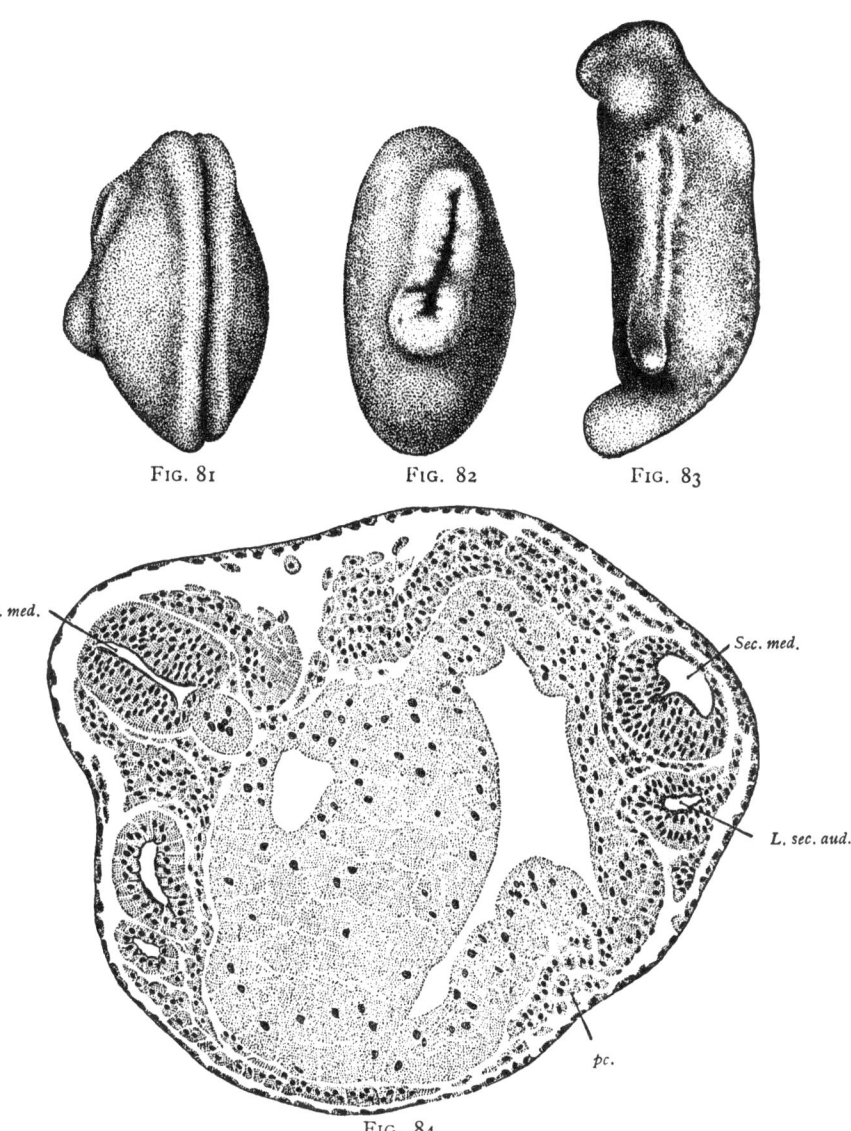

FIG. 81   FIG. 82   FIG. 83

FIG. 84

FIGS. 81 to 84.—Action of pieces of the organizer from a germ of *Triton cristatus* upon the germ of *T. tæniatus*; induction of a secondary medullary rudiment. Fig. 81, neurula of *T. tæniatus*, dorsal view; normal medullary folds formed along the whole length of the germ; to the left of the drawing the secondary medullary rudiment is seen from its right side (about 44 hours after implanting the organizer in the left side of the *tæniatus* germ). Fig. 82, side view of the same germ; secondary rudiment seen from above. Fig. 83, side view of the same germ about 5 days after the operation. Secondary embryonic rudiment seen from above, with its tail-bud, medullary tube, somites, and ear-vesicle. Fig. 84, transverse section through the embryo shown in Fig. 83 at the level of the primary pronephros; to the left and above are the primary axial organs, the secondary are to the right; *pr. med.*, primary medullary tube; *sec. med.*, secondary medullary tube; *l. sec. aud.*, left secondary auditory vesicle; *pc.*, pericardium. (After Spemann and Mangold.)

medullary tube is closed; at the front end its enlargement and the thinness of its roof are suggestive of a brain. The auditory pits are already closed vesicles, but are still connected with the epidermis; pronephric ducts, somites, and notochord are also present. Though a second gut is not indeed present, the secondary embryonic rudiment seems to have shared to a certain extent in the formation of the gut. The secondary embryo is formed partly of cells of the implant and partly of cells of the host. There can be no doubt that the development of these latter has received its orientation through the inductive action of the implant.

It is usual, as in this example, for the secondary rudiment induced by an implant to lie parallel with, and in the same direction as, the primary embryo. The host must therefore exert an influence on orientation. Whether, and in what degree, the primary rudiments of the host may collaborate in the formation of particular organ-rudiments of the secondary embryo is a question which will not here be discussed; but it certainly seems that the collaboration of the host's organizer is not necessary for the appearance of the first rudiment of the secondary embryo.

It is interesting to note that in the induction of a secondary embryonic rudiment the implanted organizer may belong to a different species, or indeed to another genus or even order. For example, a piece of the organizer from the gastrula of *Bombinator* can produce a secondary rudiment in the young *Triton* germ.

An inductive influence is exerted not only by early chorda-mesoderm material which is invaginated during gastrulation and passes under the presumptive medullary material, but also by the early medullary plate and the early notochord itself. A piece of the medullary plate of *Triton* introduced into the blastocoel of a *Triton* gastrula can cause the ectoderm of the host to form a medullary plate. We will give a somewhat fuller account of an experiment designed to test the inductive action of the early notochord.

From a young neurula of *Triton alpestris*, in which the

medullary rudiment had not begun to project, a small piece of notochord was taken, and implanted into the blastocoel of a *Triton tæniatus* embryo. Two days after the operation, the host was so far developed that its medullary folds were closed to form a tube, Fig. 85 shews the germ at this stage in side view. A small longitudinal medullary rudiment, bordered by little folds, is easily recognizable (ind. med.). Study of sections shews clearly that we have here a structure induced

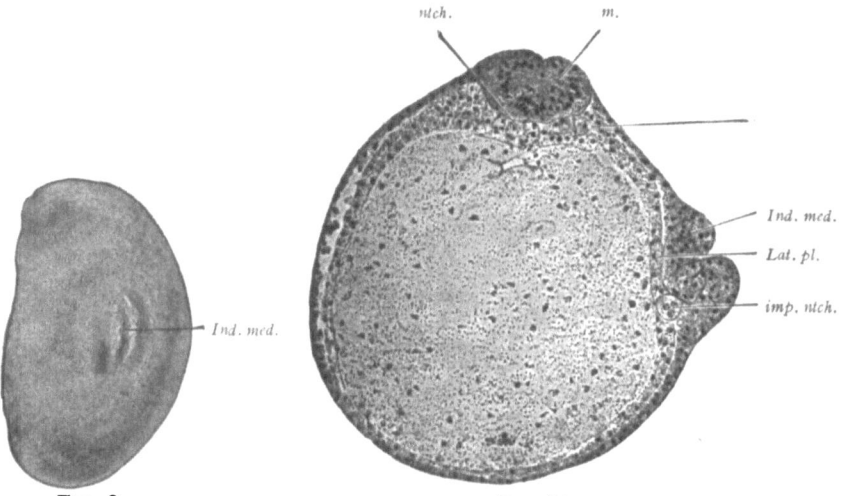

FIG. 85   FIG. 86

FIGS. 85 and 86.—Induction of a medullary plate in the germ of *Triton* by implanted notochord. Fig. 85, side view of germ two days after operation, the induced medullary rudiment (*ind. med.*) easily recognizable. Fig. 86, transverse section of the same germ; *ntch.*, notochord of host; *m.*, medullary tube; *lat. pl.*, lateral mesoderm of host; *ind. med.* induced medullary rudiment; *imp. ntch.*, implanted notochord. (After Bautzmann.)

by the implant (Fig. 86). The induced medullary rudiment is not yet completely closed to form a tube; it lies immediately above the implanted notochord, which itself is embedded in the primary lateral-plate mesoderm of the host. The development of the mesoderm of the host is not affected by the implant.

The notochord retains this power of induction for a considerable time; notochord material taken from an embryo with optic vesicles is still able to evoke a medullary plate in another embryo, though not so perfectly as can a younger notochord.

The process of determination with the blastopore as its organizing centre exists even in normal development, though there it is not so easily demonstrable as in experiments. The fact that this action may result in the production of axial organs from foreign material is of capital importance. For it follows that the cells of the dorsal lip of the blastopore are not simply the axial organs in an early embryonic stage, though they do in fact induce the development of these organs, and therefore really take part in the determination of the fate of the cells actually composing them. What really is the mode of action of the organization-centre one cannot at present say with certainty, and it is better to leave the question open for the present. The fact that the cells of the blastopore lip are turned in during gastrulation, and are ultimately used up in the formation of organs, does not affect our conclusion that this germinal region is a source of determination. This illustrates a general phenomenon; that hand in hand with the differentiation of the germ there arises a connexion between processes of determination and particular subdivisions of the germ.

Thus, the axial organs of the Amphibia do not develop completely independently, in accordance with a mosaic theory, but in relation to other parts of the embryo. Such interdependence is not an isolated phenomenon, but is of great general importance in development.

The position of the future organization-centre is not recognizable in the newt's egg before cleavage, nor in the blastula; it is not indicated by a special state of the surface of the egg, or in any other way. Not until the blastopore begins to appear can the source of determination of the axial organs be identified. This holds also for the germs of many of the Anura, as for example *Bombinator*. It is obvious that the plane including the primary egg-axis and the middle of the dorsal lip of the blastopore is the symmetry-plane of the embryo, since it is in this plane that the unpaired axial organs lie. Now, the same orientation holds for the grey crescent of the egg of *Rana fusca*. This represents the presumptive material of the dorsal lip of the blastopore, so that it must be regarded as presumptively equivalent to the future organization-centre. The difference between the germs of *Rana* and of *Triton* is simply that in *Rana* the centre has already begun to appear before cleavage, while in *Triton* it becomes apparent only after cleavage.

The grey crescent can be regarded as analogous to the pole plasms in the eggs of other forms of animals. The substance of these also is used for the formation of organs it is true, but their significance does not end there; though morphologically quite different, pole plasms resemble the dorsal lip of the blastopore, which arises from the grey crescent, in being at the same time organizers and inducers.

Whether in all embryos which form a typical blastopore lip, this lip always plays the part of an organizer is a question which must remain for the present undecided. For example, the results of experiments on Echinoderms partly support and partly are against this view, so that further results must be awaited. The question also arises as to whether there is not perhaps always present in the germ a special subordinate organization, which can be called an organization-centre, even when no typical blastopore is formed. In the case of Insects, for example, certain observations seem to point to the presence of such a centre at the hinder end of the very young germ, but other facts cannot at present be brought into accord with this. The general view that the fundamental processes of development are everywhere the same certainly favours the idea that, ultimately, in all animals a subordinate organization will be found which acts like the blastopore lip of the Amphibia, i.e. as the principal source of organization and determination. We can assume also that its condition, morphological and other, and the time of its appearance may vary very much with the individual. Since these conditions are already very well understood in the case of Amphibians, their explanation in the wide range of other forms can only be a matter of time. The theoretical significance of the phenomena established in Amphibians is in no way prejudiced by the foregoing reservations.

Consideration of these phenomena of induction might give the impression that development was based upon the independent action of the parts, and that the whole germ was merely the sum of such active parts. If this were true, the formation of the axial organs could not take place without the participation of the blastoporal organizer. It can be shewn, however, that axial organs such as the notochord and the neural tube can arise from early germinal material even when the action of the organizing region is completely eliminated. A number of different types of experiment witness to the truth of this fact.

For example, the presumptive material of the axial organs can be freed from the possibility of influence by the organizer by removing this material—or part of it—from the germ before the formation of the blastopore, that is to say in the blastula stage, and by then culturing it in isolation. Other agencies conditioning determination and subsequent differentiation must exist apart from the organizer-region in question, if the formation of parts of the axial organs still occurs in these conditions.

The culture of such isolated parts of the blastula can be carried out by the method of interplantation (p. 43). It is possible in such interplants to observe the development both of the notochord and of the medullary tube, and to culture these organs—or at least parts of them—completely isolated. Axial organs can arise either from interplanted parts of the blastula or of the young gastrula. In the former the material has not yet come under the influence of the blastopore organizer, since this region is not yet formed; the organs must therefore have arisen without its inductive action. This is sufficient to prove our original contention—that other agencies than the organizer-region of the blastopore are concerned in the determination of these organs. But further, it follows that in such interplants notochord and neural tube can develop not only from presumptive chorda- and medullary materials respectively, but equally well from materials of other presumptive organ-regions which never normally produce these organs. Although it is not always possible in the blastula—owing to the absence

of distinctive features—to determine the position of the blastopore, and therefore that of the presumptive germ-regions, yet
the place of origin of the interplant can always be exactly
defined by waiting for the gastrulation of the donor; for when
it occurs a defect caused by the excision always declares itself,
and the position of this makes it possible to assess the normal
presumptive value of the interplant. Obviously what is interesting in this connexion is the appearance of axial organs from

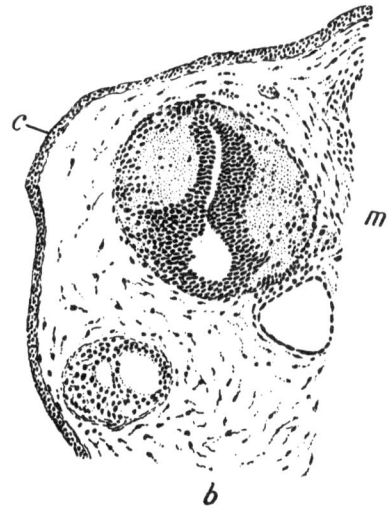

FIG. 87.—Differentiation of neural tube in an interplant
of presumptive epidermis of *Hyla arborea*. *c*, conjunctiva
of host (a larva); *m*, medullary tube—developed from the
interplant—seen in cross-section; *b*, connective tissue of
host. (Original from a preparation by W. Kusche.)

material which normally produces quite different structures.
For such material cannot possibly have received either from
the blastopore lip or from its cytoplasmic precursor an impulse
inducing the formation of axial organs.

Excision of materials from the early gastrula can be carried
out with greater precision than from the blastula, for the
topography of the presumptive regions of the gastrula is known
with some certainty. Fig. 87 gives the result of such an experiment. In a young gastrula of the tree-frog, *Hyla arborea*, a

definitely circumscribed piece of presumptive epidermis was excised from the side opposite to the blastopore, and interplanted in the orbit of a young tree-frog larva, from which the eye had been removed. The interplant is seen to have differentiated as a piece of neural tube which shews a completely typical arrangement of its elements. In the section illustrated it exhibits clearly some of the characters of a brain—the shape of its cavity (= ventricle), the thinning-out of its roof (below in the figure), the relative positions of ganglion cells and fibres. Indeed it corresponds in its general character fairly well to the condition of the larval thalamencephalon, though, as we should expect in such a case, some few irregularities occur in its configuration.

That the formation of the medullary plate is essentially independent of the presence of underlying chorda-mesoderm material is shewn in a differently planned series of experiments, in which the material is either entirely removed before gastrulation, or is prevented from assuming its normal position.

Though these experiments shew that induction by the organizing region in question is not solely responsible for the determination of axial organs, the importance of the lip of the blastopore in normal development remains unaffected; for on it normally depends the *destination* of the axial organs. In their *institution* at least other forces are obviously concerned, but these as a rule seem to take no very important part, their intensity being less than that of the induction of the organizer. In certain circumstances, when the organizer is eliminated, these secondary agencies suffice to control development in the cell-complexes up to the time of their destination and consequent differentiation. The conclusion is thus reached that more processes than one are concerned in the determination of the axial organs of the Amphibia, but that one of these —an influence radiating from the blastopore lip—is really decisive, while the others are concomitant. In cases of necessity, however, these latter can alone produce the result. In the determination of the axial organs we see, then, a phenomenon which may be called *double assurance* ("making doubly sure"), or to use a more general term *multiple assurance*. It might be even better to speak of multiple causation. Such multiple assurance or causation of a deyelopmental process rests upon the fact that all the cells of the germ have the same specific reaction-basis, and upon the other fact that the germ is not a mosaic of heterogeneous parts but a unit and a whole. Development accordingly depends not upon any special part alone, but upon the whole germ, though a special part—as in the case of the blastoporal organizer, or of the influence proceeding from it—may have predominance. Development is thus ultimately an integral process.

III. Determination as a Dependent Phenomenon in Later Stages of Development and in Regeneration

1. Organ-Rudiments as Organizers

Organization-centres are by no means the only sources of determination which come into action in the course of development; but they are the first to appear as organizers. Later they are associated with organizers of a higher degree.

If we designate the blastoporal organizer as one of the nth degree, those which later appear may be said to be of the $(n + x)$th degree, those which are perhaps earlier active in the germ, of the $(n - y)$th degree—where x and y may have the most varied values. Organizers of relatively higher degree either have themselves been induced by those of a relatively lower degree, or—in the case of organizers of the lowest degree —they depend directly upon inherited factors. The degree to which organizers belong is decided, however, not merely by the chronological sequence of their appearance, but by the more general or more limited character of their activity; for, here again, it appears that there is a progress during development from the general to the particular. The organizers, on account of the difference in their substratum, are themselves different in size before and after cleavage of the egg; at the same time they take on, during development, more and more the character of organ-rudiments, and thus depart from the character of the first organizers. Possibly in the future we shall have, on this account, to distinguish completely different categories among them. For the present, however, we may be content to speak quite generally of organizers.

We have an organizer of the higher degree in the case of the embryonic optic cup, the influence of which is essential in the determination of the lens of the eye. Now the optic cup itself does not arise as a self-differentiation, directly preformed by genes. At an early embryonic stage it is replaceable, and can be formed, in that case, from foreign material. This simply means that the optic cup is induced again by an organizer,

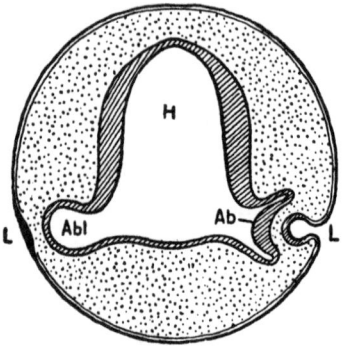

Fig. 88.—Diagram to explain the mode of development of the vertebrate eye and its lens. The figure—a transverse section through the head of an embryo—represents diagrammatically an earlier stage on the left side than on the right. On the left the primary optic vesicle (*Abl*) has grown out from the brain (*H*), and the lens (*L*) has appeared as a thickening of the ectoderm of the head. On the right the optic vesicle has become the optic cup (*Ab*), and the lens-rudiment has become sac-like.

though in normal development the presence of this organizer may not be apparent.

It is well known that in vertebrates the eye, or rather the primary optic vesicle, grows out from the embryonic brain towards the surface of the head (Figs. 88 and 89). The outer wall of the vesicle then sinks in, and the rudiment assumes the form of the optic cup. Where this approaches the outer epidermis a sac is invaginated from the latter towards the interior of the cup. This sac is ultimately cut off completely from its place of origin, and, as a rounded vesicle surrounded by the optic cup, becomes the rudiment of the lens. Experiment has shewn that the formation of the lens is dependent on the development of the optic cup, and, indeed, that the latter exercises an inducing action on the ectodermal epithelium of the head.

Both extirpation and transplantation experiments have con-

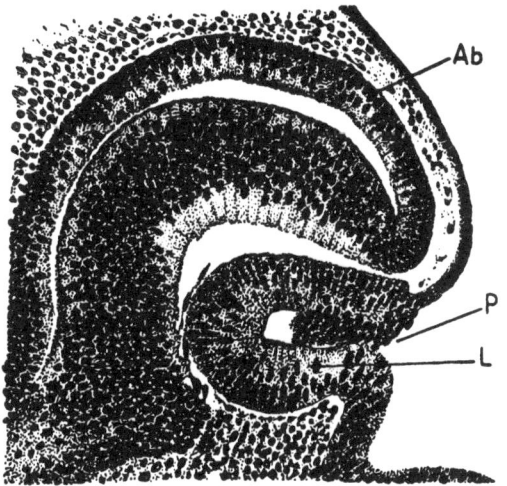

Fig. 89.—Section through the developing eye of an embryo pig. The section is in the plane of the choroid fissure (below in the figure), so that the optic cup is here incomplete. *Ab*, optic cup, double-walled as a result of the indentation of the original optic vesicle; the inner layer—already greatly thickened —becomes the retina; *L*, lens-sac still in connexion with the head epidermis; *P*, plug of cast-off and degenerating cells (characteristic of the Pig, but absent in most animals). Connective-tissue cells are pushing in from below, between the lens-vesicle and the optic cup—these take part in the development of the vitreous humor.

tributed evidence of this action. When, in the Anuran embryo, the presumptive rudiment of the embryonic optic cup is extirpated before the appearance of the lens, the development of this last does not, in the great majority of cases, take place. The operation is best carried out by excising, in the neurula stage, that part of the neural fold from which the primary optic vesicle later arises (Fig. 90). In this manner the formation of an optic cup is prevented, without injury to the part of the presumptive epidermis from which the lens normally arises; for this part still lies ventral to the neural groove, on the lateral surface of the embryo. Only after the coalescence of the edges of the neural tube to form a median-dorsal suture is that part of the epidermis drawn dorsalwards, to lie against the outer side of the embryonic brain. If, then, that part of the medullary plate which gives rise to the fore-brain (and therefore to the optic vesicle) be previously cut out, any influence upon the region of the lens-rudiment is eliminated. The failure of the lens to develop after this operation implies its causal dependence on

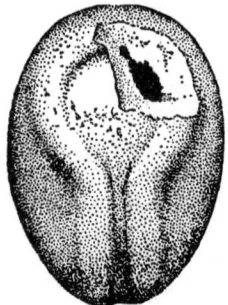

FIG. 90.—Neurula stage of *Rana esculenta*. The right anterior half of the brain rudiment, from which is developed the optic vesicle, has been cut out. (After Spemann.)

the action of the optic cup. A lens does not, indeed, in every case fail to develop, but if a "free" lens is formed without an optic cup, it can never achieve complete differentiation. In all species of Amphibia, at any rate, there seems to exist the fundamental ability to form a lens without the action of the optic cup; and, in respect of the frequency of formation of such free lenses, differences seem to exist among the different species. Nevertheless it is certain that the causal relation in question is present in normal development; for the dependence of lens-formation on the action of the optic cup can be demonstrated in another way—and positively—by presenting to the optic cup foreign material out of which to produce a lens.

For example, it is possible, without injuring the optic cup, to replace—by a suitable piece of epidermis from the trunk—that part of the head-epidermis from which the lens is formed (Fig. 91). If this operation be performed upon the embryo of *Rana esculenta* shortly after the closure of the medullary folds, a lens is formed from the foreign transplant. The optic cup does not, however, produce a lens from trunk-epidermis which is relatively older. The same experiment on *Bombinator pachypus* and *Hyla arborea* has given similar results. It is possible to transplant the primary optic cup under another part of the epidermis of the same embryo, in which case the transplant will often cause the formation of a lens from this

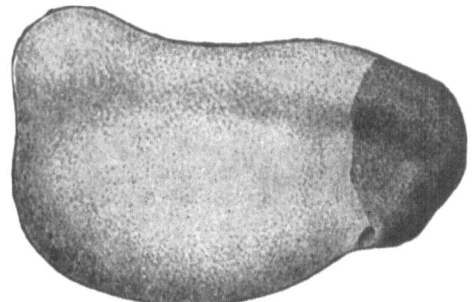

FIG. 91.—Early embryo of *Rana esculenta* in which the normal head-epidermis of the right side has been replaced, before the formation of the lens, by transplanted skin from the trunk—recognizable by its darker pigmentation. (After Spemann.)

foreign epidermis. Experiments on Urodeles and Fishes confirm the dependence of lens-formation upon the influence of the optic cup, though in certain circumstances, in these animals as in the Anura, free lenses may arise.

We may sum up thus. There passes out from the optic cup to the overlying epidermis an influence which decides the fate of the epidermal cells. The formation of the lens is dependent upon the influence of the primary optic cup; and this may also be assumed to be the case for normal embryonic development in those forms where, in experiment, a lens or lens-rudiment can arise even in the absence of an optic cup. At the same time, the degree of dependence may vary somewhat in different species. Since, however, a lens may sometimes be formed without the optic cup, there must be at least one other factor—perhaps several—participating in the determination of the lens, and possibly participating to different extents in different species, according to whether the influence of the optic cup is greater or less. We have here again cases of what we have called multiple assurance or multiple causation. This other factor, however, plays no very prominent part in normal development, in which it is the influence of the optic cup that principally activates the potentiality to form a lens; but it can, in certain circumstances, produce the lens unaided.

The eyeball in Vertebrates, later in development, also behaves as an organizer. The conjunctiva forming the corneal epithelium (which, arising from the epidermis of the head, finally fuses with the surface of the eye) depends for its specific differentiation upon the presence of the underlying eye. That part of the head epidermis out of which the corneal epithelium is formed in Amphibia is at first pigmented like the rest of the skin. If the eyeball is extirpated early, without damaging this region, de-pigmentation and the other differentiations peculiar to the corneal epithelium are arrested, and it remains in exactly the same condition as the rest of the head epidermis. A positive proof can be given of this influence of the eye on the epidermis lying above it, by transplanting the eye, or parts of the eye, under the trunk epidermis. In larvæ of *Salamandra maculosa* the skin over such a transplant is greatly changed; the unicellular glands (Leydig cells) degenerate and assume the same condition as the rest of the cells of the epidermis; the epithelium becomes two-layered, clearer, and in general very like a normal corneal epithelium.

In larvæ of the frog, differentiated corneal epithelium can be removed, and replaced by a transplant of differentiated epidermis from the body. In a comparatively short time the transplant is converted into typical corneal epithelium. Such experiments prove that the influence of the eye does not merely assist in the formation of a corneal epithelium which has been otherwise determined, but that this influence directly affects the actual determination of the cell regions in question. For upon the influence of the eye depends the specific quality of the differentiation. We may therefore claim, with regard to the formation of the epithelium in question, that the eye is an organizer.

Such observations point to the existence of manifold inter-dependencies among the parts of the developing embryo. It has been said that development is a function of the whole, but that in the course of development many of the organizing activities of subdivisions of the germ are lost. We must now emphasize the fact that the special importance of the parts diminishes still further with the progress of development. When all the differentiations are completely determined, interdependence of the parts disappears progressively, so that in the final stages of development the parts no longer possess an organizing function. It need hardly be said that the time at which the parts become independent differs greatly, according to whether the processes of determination are much drawn out during development or are crowded into its early stages.

2. Organizing Action in Regenerated Parts

Organizing activities occur also in processes of regeneration. Apart from an organizing influence exerted by the residual portion after an experimental amputation, it is quite impossible to explain the fact that the new formation exactly replaces the lost part. These processes of induction would seem to be especially necessary where the regenerate arises from an extensive blastema which primarily consists of more or less undifferentiated cells, but in any case does not contain, pre-formed within it from the beginning the tissues and morpho-logical regions of the future regenerate. It is to be noted, in fact, that the regenerate is not formed, in such cases, by pro-liferation of the parts which remain, but that it represents actually a new rudiment within the blastema. It is possible that the inductive action of organizers may be quite general in regenerative processes; in particular cases its action is patent.

Since we are here concerned only with the fundamental facts, a few examples will suffice. They will shew, among other things, that the activity of organizers, though in embryonic development it is no longer noticeable after determination is completed, is by no means entirely absent in the fully differentiated organism and in its parts.

Investigations into the polarity of transplants provide a

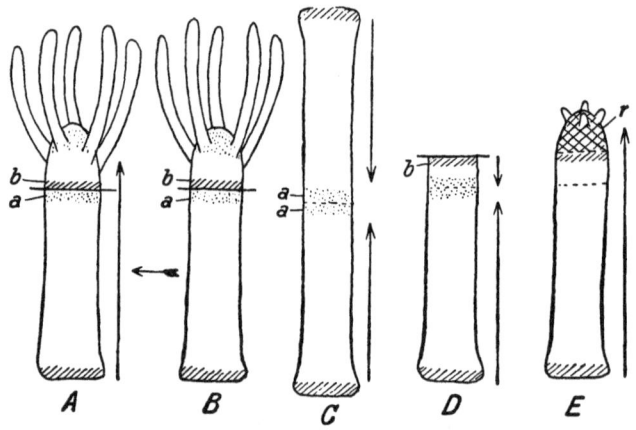

FIG. 92.—*A–E*, diagram to explain reversal of polarity in the regeneration of a part of *Hydra* forming an inverted transplant; fusion of two apical cut-surfaces, and regeneration after transection of one of the fused components. Arrows indicate polarity. *a*, apical (oral) cut-surface, *b*, basal (aboral) cut-surface; *r*, regenerate.

fairly simple case of organizer-action upon regenerating parts. Polarity in transplants is generally strictly maintained; but it may, during regeneration, undergo a reversal which is obviously due to a definite influence from the bearer of the transplant.

Valuable results have been obtained by joining two components, each of which has a strongly marked polarity, by their similar poles. This succeeds best in the case of worms and polyps. If in two individuals of *Hydra* the tentacle-bearing, apical ends are cut off (Fig. 92, *A, B*), and the two basal pieces are made to grow together, then the two components of this combination are joined by their similar (apical) cut-surfaces (Fig. 92, *C*). This is most simply performed by threading the pieces upon a bristle; they at first contract, but that is an advantage because their subsequent extension brings the surfaces of the wound all the more closely together. This "transplant-combination" possesses no mouth-opening, but has a basal plate at either end. If now the one component be cut through in the neighbourhood of the place of junction (Fig. 92, *D*), it thus acquires a free basal cut-surface. From this last there is not regenerated, as one would expect, a basal end, but a mouth-end with tentacles—in opposition to the original polarity (Fig. 92, *E*). As a further result of such experiments it has been shewn that even a small piece of a *Hydra*— which on account of its minute size is itself no longer capable of development—can alter the differentiation of much greater masses and force them to take its own direction in development.

In regeneration it often seems that the kind of regenerate formed is influenced by the regenerating individual as a whole. For example, if half the head-end or half the tail-end of a Planarian be removed by a cut at right angles, the regenerate produced will vary in kind according to whether the cut lies farther forward or farther back; that is to say, according to how much of the operated half of the body is left. What the causal connexion may be in such cases is not yet, however, fully understood. From this point of view the phenomena of regeneration are somewhat clearer in the case of the Amphibian limb.

In the regeneration of a foot of *Triton* it is not the whole body which influences the quality of the regenerate, but only the stump that is left after amputation. For, if in the adult *Triton* the fore-limb and the hind-limb are exchanged by transplantation, and then amputated after healing in, the transplanted extremities always regenerate in accordance with their origin; so that the character of the regenerate is not due to induction by the whole body. That such an influence is exerted by the remainder of an organ can be shewn, however, in various ways—for example by observing what happens in regeneration from a duplicated cross-section. In *Triton* this can be done by bending the upper arm and fore-arm at the elbow, so that they lie parallel, having first skinned the radial side. Thus the two parts of the arm become fused side by side. By amputation of the elbow a single wound-surface can now be made, to which both the upper arm and the fore-arm contribute. To outward appearance there is a single regeneration-blastema, but the regenerate arising from it is bilaterally symmetrical. From this it follows that both components of the cut-surface have exerted their influence, but in an enantiomorphic fashion corresponding with their structure and spatial relation to one another. In other words, the remainder of the organ has interfered in the determination of the regenerate.

The regenerate, however, is not determined in detail, part for part, by the cross-section of the remainder of the organ. This is proved, first, by the fact that half a cross-section of a limb can produce a whole regenerate; and, second, by the further fact that if a femur be substituted for the humerus, a typical hand and not a foot is regenerated, after healing, from an amputation-surface in this region. The new bone arises free in the blastema, without connexion with either the bone of the stump or with that which is introduced. We have already seen how a limb-stump free from bone can form a regenerate possessing a typical bony skeleton. On these grounds one is forced to limit the influence of the remainder of an organ to an action of orientation affecting the whole blastema.

This general action of orientation, as it must provisionally be called, clearly contains a specific component, since it may also affect the quality of the regenerate. The following experiment is of great interest from this point of view. If a piece of the tail is amputated in an adult newt, the animal forms at once a regeneration-blastema, which when left in its normal place regenerates a tail. If, however, such a "tail-blastema" be transplanted into the neighbourhood of the base of the fore-limb, varying results are obtained (Fig. 93, *a–c*). Apart from cases in which the transplanted tail-blastema does not develop further, it may yield a little tail by differentiating as if in its place of origin (Fig. 93, *a*), or, conversely, a well-formed limb by the opposite process. In this case an accessory limb is

formed from tail material, near the normal fore-limb
(Fig. 93, *b*, *c*). That this limb has, in fact, arisen from the tail-
blastema, and not from trunk tissues of the host, is seen from
the following fact: the accessory limb is not in close union
with the tissues of the host, but in an advanced stage has
almost completely separated from its substratum, to which it
remains attached only by a thin thread of skin (Fig. 93, *c*).
Where the regenerate develops in accordance with its place

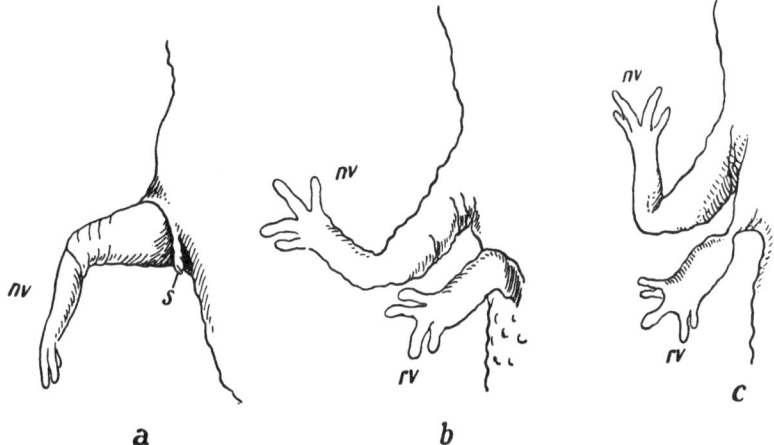

FIG. 93.—*a–c*, development of a regeneration-blastema of the tail of *Triton*
transplanted in the neighbourhood of the fore-limb. *a*, differentiation in
accord with original situation: formation of a minute tail (*s*); *b*, differentiation in accord with new situation: formation of a typical limb; *c*, later
stage of *b* illustrating the fact that the regenerate is independent of the
tissues of the host. *nv*, normal fore-limb; *rv*, regenerative limb. (From
original sketches by P. Weiss.)

of origin, this can only be explained by supposing that the
transplanted tail-blastema was already determined. To account
for a regenerate of the opposite kind, we must conclude that
the blastema at the time of transplantation was still undeter-
mined, and was subject to organizing or determining influences
of the new environment. Before further conclusions can be
drawn from such experiments a more thorough investigation
must certainly be made. But the existence in regeneration of
activities which essentially resemble those of an 'organizer'
—though in any case it can be deduced from a study of regenera-

tive processes—receives, in the experiments just described, a particularly convincing proof.

IV. THE EPIGENETIC CHARACTER OF DEVELOPMENT

From the results of investigating the potency and regulative capacity of the parts of the germ, we have, in an earlier section, deduced the non-preformational character of development. We can now go a step farther, and amplify this merely negative characterization of the process of development by making the definite statement that it is essentially epigenetic. The grounds for this statement consist in the indirect relation borne by the original rudiments to the ultimate differentiations of the living organism, and in that principle of progressive organization which dominates development.

According to the theory of Preformation nothing new arises in development; the same diversity is present in the fertilized egg as is found in the developed organism. This latter contains a great number of morphological structures and differentiations; these will be present in the germ in its initial unicellular stage, in a more or less encysted (*eingekapselt*) condition, since for the direct preformation of every structure a discrete rudiment exists. According to modern preformist views these discrete rudiments consist simply of corpuscular constituents of the chromosomes. Development is a mere unfolding of these rudiments, which are directly related to those particular differentiations of the complete organism which they represent, even though the material corpuscles produce the definitive structure of organs and tissues only as the result of a long chain of reactions.

In contradistinction to this, the conception of Epigenesis involves the production of new manifoldness during development. The individual differentiations of the fully-formed organism are in no way represented by a corresponding number of preforming structures at the beginning of development; but the more complicated condition follows gradually from one which is really simpler. The presence of rudiments, and even of material rudiments, in the original unicellular stage, is indeed assumed. But these rudiments do not directly correspond, either in number or in character, to particular differentiations finally present: the relation is only an indirect one. The rudiments too are not sufficiently numerous to fulfil the requirements of a preformation theory; nor are we concerned solely with discrete corpuscles, lodged mosaic-like in the chromosomes: the whole constitution of the basis of development or, as we have called it, the reaction-basis, plays an important part. Development is not a simple unfolding, but a real origination —in the first place the origination of factors of development which only shape themselves little by little into the final individual factors with which particular differentiations of the organism are connected. These final factors have, then, a relatively direct relation to the tissue- and organ-differentiations belonging to them. The relation, however, between the rudiments initially present and these definitive differentiations of the developed organism is not only a *mediate* one—as it could also be according to Preformation—but is really indirect.[1]

There may, indeed, be present in the unicellular stage of the organism some rudiments which are directly related to later differentiations and characters of the developed animal; but these do not really control development as a whole, they can only play an ancillary part in connexion with structures of lesser importance.

In view of the importance of the questions concerned we will restate the matter as follows. By the "direct" relation of a rudiment to a character or differentiation must be understood the actual preformation of the peculiarity in question— whether it be essential or accessory—by an actual concrete corpuscle in the germ-plasm or in a chromosome, exactly in the sense of the mechanistic, "evolutionary" theory of Preformation to which the modern chromosome theory of heredity belongs. The realization of such a rudiment in terms of the developed character may conceivably occur in one or other of two ways. The first possibility—generally denied at

[1] See Translators' Preface.

the present day—is that the rudiment is itself *immediately* converted into the organ or distinctive character in question; the second is that the rudiment sets in motion a series of physico-chemical reactions the course of which leads directly to the specific character. This latter method of realization is indeed mediate in so far as it is not the rudiment itself which is converted into the character—since there intervene a great many intermediate steps—but the connexion between rudiment and character is nevertheless direct, that is to say preformational. The whole development of the embryo consists, then, of concurrent series of such reactions leading directly to their end. A really new diversity does not arise in this way: the components of the diversity already present merely change their individual character.

By "indirect," as applied to the relation of a rudiment to a differentiation of the completed organism, it must be understood that the differentiation, as such, is in no way really preformed in the rudiment. The number of discrete rudiments at the beginning does not therefore correspond to the number of differentiations later to appear, but is smaller. Thus the original condition of the germ is actually simpler than its final condition after the completion of embryonic development. The action of a rudiment initially present is not the creation of any one character or differentiation, but the changing of a relatively simpler and lower initial organization into one which is relatively higher. This last, by virtue of having acquired new organizers, now acts in the same way, so that finally a greater complication is produced than existed at the outset. Development is not preformist Evolution, but Epigenesis. This "indirect relation" is naturally at the same time mediate, although obviously in a different sense from that in which a direct relation may be mediate. There is not present in the germ-plasm a discrete rudiment corresponding to every peculiarity of the developed organism.

The first reason for believing that the primordial rudiments have only an indirect relation to the definitive characters of the organism is based on numerical considerations. If development depends upon the preformation of all the parts, there must be present in the zygote at least as many separate material rudiments as, later, there are differentiations. It is further to be observed that, from the standpoint of Mendelian genetics, several—indeed many—genes are very often concerned in the formation of a single character. Again, the individual genes are said to be in very many cases multiple, that is that they are again subdivided into a large number of similar parts. If we bear in mind—what no hypothesis of preformation can avoid assuming—the immense number of essential rudiments united to material parts of the germ-cell, we shall see that this number is much too great to make it possible that all these corpuscles should find room in the germ-cell, and, *a fortiori*, in its nucleus. If this is true for the egg-nucleus, it applies with even greater force to that of the sperm, which is generally very minute. For this reason alone the relation of character to rudiment cannot be such that each essential and each accessory character is directly represented in the zygote by a special material gene. It follows from this that pure preformation is an impossibility, and that we must assume a relation between rudiment and character which is only indirect—in fact, the epigenetic character of development.

The impossibility of a direct relation between rudiment and final differentiation follows also from the facts of regulation. Complete preformation, or, what is the same thing, a direct relation between the rudiment and its result in development, would mean that the mechanism of development could only work along definitely fixed lines; any disturbance must bring catastrophe, and regulation would be out of the question. That, however, is certainly not the case. Nothing remains for us, then, but to ascribe not only a mediate but an indirect character to the relation between the primary factor and its material carrier on the one hand and the differentiations of the developed organism on the other. In other words again, development is essentially epigenetic.

For the rest, it is not only discrete corpuscles—contained in definite morphological structures of the germ-cell—which are to be regarded as factor-carriers: the specific character of the living protoplasm as a whole is definitely of outstanding importance. This formative constitution of the protoplasm, however, can only indirectly condition the detailed differentiation of the organs which appear later, because it does not contain a definite region for each character.

Now experimental embryology also shows that the factors of the zygote do not lead by any means directly to the definitive characters, but that they condition these only indirectly; for at first there are called forth only those preliminary steps in differentiation which we have called above subordinate organizations, and these in their turn become responsible for further development. These phenomena exemplify the principle of *progressive organization*—to use Spemann's appropriate expression—or as we might call it, progressive and gradually specialized determination.

If we recall the origin of the axial organs in the Amphibian germ, and in particular the neural tube, it is at once obvious that the determination and differentiation of this last are not due to a gene working directly to that end, but that the organ in question is only indirectly connected with the complex of rudiments. First from certain rudiments is formed the lip of the blastopore (ultimately traceable to a particular cytoplasmic differentiation of the egg); from this subordinate organization, which acts as an organizer or inducer, proceeds the induction which determines the medullary plate. We have seen, further, that this organizer is by no means the only one of its kind—indeed, on the contrary, that wherever the determination of a structure is in progress, owing to the influence of one part of the germ upon the rest, we have to do with an organizer of one degree or another.

Thus, development is not the mosaic-like unfolding of directly preforming genes, but the creation of a whole progression of organizers. These, on the one hand, give rise to further organizers of a higher and more specialized order; and, on the other hand, they themselves accomplish the definitive determination of the tissues and organs. Hence, the realization of the thing determined is again conditioned by manifold activities which are not directly represented in the mass of rudiments, but are only made possible as the result of a gradually increasing variety of organizers and primary differentiations. In the gradation of organizers there is a progression from the generalized to the more specialized.

Epigenesis gives to the process of development much more freedom than could preformation, for it does not preclude regulation and regeneration. Difficulties presented by the demand for too great a number of corpuscular rudiments also disappear. What alone is preformed is the specificity of development as a whole and this by means of the specific reaction-norm of the whole reaction-basis. But the details of differentiation are not preformed by the presence of individual genes directly representing each special structure. The origin of individual differentiations is, again, epigenetic, since the relation of the first condition to the final product of development is only indirect. The reaction-basis of the new individual is of course inherited. What is inherited is not simply discrete carriers of individual characters, but a whole reaction-basis containing definite factors which react in a definite way with the internal and external factors. By this means progressive stages in organization are created, which give rise to factors more and more specialized, and these finally realize the external characters. These specialized factors stand, therefore, not at the beginning but at the end of development, and, for this reason, they can be called final factors. They have a direct relation to the characters of the phenotype. Since they produce accessory racial characters, and to some extent specific characters also, they can be studied in crossing experiments.

In connexion with the phenomena of potency and regulation it has been pointed out that two types of development can be distinguished—the regulative type and the mosaic type. Now, it might appear that the foregoing discussion would apply to the regulative type but not to the mosaic type; because in the latter we find that a high degree of independence of the parts is already present at an early stage, so that inductive activities are not noticeable. It must be borne in mind, however, that at least some of the relations which earlier led us to deduce the epigenetic character of development are of a fundamental nature; so that one would expect to find them also in the mosaic type, though perhaps in a somewhat different guise. Independence of the parts in development and lack of interrelations can be deceptive, by reason of dependence having been present very early; self-differentiation in that case only follows upon a previous dependence. We are thus dealing with a very precocious occurrence of determination; and the fact that the decisive moment in development comes at different times completely explains the contradiction between the regulative and the mosaic types, as we have already said.

While in regulative eggs the gradual progress of organization and determination extends over a great part of embryonic development, and in the majority of cases begins only after the cleavage of the egg, these processes are crowded together into a much shorter space of time in the extreme mosaic type, and generally occur before cleavage. In this case the final factors mentioned are formed very early. But it must be remembered that the division of the germ into cells does not in the least alter the fact that it forms an integral protoplasmic system. The amount of crowding together of progressive determination varies, and all possible transitions occur in this respect. Once the determination of a process is complete, development takes on the appearance of preformation. If, then, determination may occur either earlier or later as the case may be, that does not mean that the inherited rudiments stand in one case in a direct, in the other case in an indirect relation to the definitive differentiations. It means, rather, that the times of appearance of the various stages of organization are able to vary because they have no essential importance. In a word, there is no reason to doubt the truth of the general statement that development is epigenetic.

CHAPTER VI

THE INFLUENCE OF THE ENVIRONMENT ON THE PROCESSES OF DEVELOPMENT

I. THE ACTION OF EXTERNAL FACTORS ON THE INDIVIDUAL

1. General Action of the Environment

The factors concerned in development can be divided into two groups, according to whether they are connected with the inherited reaction-basis and its effect, or whether their influence upon the course of development comes from the environment. Internal factors are, of course, those which define the direction of development, and which take part in the actual events of determination. The course of development, which follows the specific reaction-norm of the reaction-basis, is influenced, however, by the external factors of the environment of the germ. It is modified by the nature and intensity of these external factors; and differences in the product of development which are attributable to dissimilar external factors are on this account called modifications. In so far as only the immediate product of development is affected in this way (and it is always so affected) we can speak of an immediate or *individual* action of the external factors; we shall see later that, in certain circumstances, there can also be an action which goes beyond the individual—a *trans-individual* action.

The environment of the germ and of the organism in general is, of course, very complex; it consists of many separate factors, each of which may be of importance in development. Exact investigation of its action can only be carried out by testing the factors separately; but that the environment is of importance in the growth of the organism is very clearly seen from its general effect. For example, if wild animals are removed from their natural environment into conditions of domestica-

tion—that is, kept by man for any purpose in artificial culture—
they shew many divergences from their normal condition, as
can be easily seen by comparing them with their wild kindred.
Domestication removes animals, though it may be uninten-
tionally, into an environment which differs very considerably
in every case from the natural surroundings. At present it
cannot be said which particular factor is responsible for the
modifications which appear. The conditions of domestication
must be taken as a complex whole, and we must be content to
affirm that there is a general effect of the whole environment.
But the fact that the external factors do have an influence on
the result of development is itself of the greatest importance.
For it follows from this that the processes of development do
not proceed automatically, within limits and along lines marked
out by the internal factors; but that on the contrary internal
and external factors act reciprocally.

The conception of domestication, as it is set forth in the
above sentences, originally had a somewhat restricted content.
It embraced only the raising of domestic animals and the
cultivation of useful plants. It was concerned therefore with
intentional domestication. The progress of human civilization
and culture has meant that far-reaching and unforeseen influences of domestication have been brought to bear on nature,
and man himself has been especially subject to such influences.
The vast alterations in the earth's surface, in the plant and
animal world, wrought by man; the development of means of
communication and the technical achievements of civilization,
which have invaded "natural" conditions; the ever-increasing
cleavage between the conditions of human life and the original
"state of nature"—all these are profound modifications of the
"normal" environment: they are, in fact, domestication in the
wider sense of the word. Though we are still far from possessing an adequate knowledge of these influences, we may be
sure that organisms, and man himself, are not unaffected by
all this. It will be a difficult but important task for the future to
investigate the significance of domestication, in this wider
sense, as a source of the modification of organisms—and not

least of man himself, who in this sense is the most domesticated of all living things.

2. Factors of the Environment Separately Considered

(a) Gravity and centrifugal force

A factor of the environment to which, from its very nature, everything must be subject is gravity. It acts first upon the organism of course, as upon all bodies, simply in the physical sense. That does not, however, exclude the possibility that it may, at times, have also a specific influence on development. Indeed, the view was formerly held that everything alive bore the impress of gravity, and that this force dominated the organization of living things. This is certainly not the case. The importance of gravity in development, particularly in animal development, is not great. Its influence is chiefly seen in the orientation of free-swimming eggs in which there is an unequal distribution of yolk, as in those of the Amphibia. Here the result is that the yolky, vegetative pole normally faces downwards. Again, the yolk-distribution in large yolky eggs is in relation to gravity, which "arranges" the substances of the egg according to their different specific gravities. The type of cleavage is largely dependent on the amount of yolk in the egg, and on its arrangement (one need only call to mind here extreme telolecithal eggs with discoidal cleavage). Hence gravity, affecting as it does the distribution of yolk, comes to have, in some sort, an influence on development, though as regards cleavage this is indirect and not specific or differentiating. The part played by gravity in the arrangement of the substances of the egg is easily shewn by turning a frog's egg vegetative pole uppermost, and forcing it to stay in this position. There at once appears a down-streaming of the contents of the egg, due to gravity. The course of development is not necessarily in any way affected by this. In certain circumstances, it is true, the production of such displacements of material is a means of modifying development; but this modification is not the specific result of the changed direction of gravity, but the indirect result of disturbances arising out of these displacements.

In advanced stages of development gravity has, in some animals, a special action in determining the direction of growth. This is so in the case of Hydroid colonies. These colonies are attached to the substratum by means of root-like outgrowths called stolons; branches of the colony, bearing the polyps, grow freely upwards. The colonies shew geotropism as it is typically exhibited by plants. In certain circumstances the polarity of the polyp colony can be reversed by the action of gravity. If a piece cut from the stem of *Tubularia* be hung with the apical pole downwards, hydranths are regenerated on both the cut-surfaces, one being the basal surface which really should form stolons. In a corresponding experiment with another polyp, *Antennularia antennina*, it was observed that in a vertically suspended piece of the stem the end that was turned upwards always regenerated buds with hydranths, while the lower end always regenerated stolons, even when the piece of stem was hung upside-down, with the apical end downwards—a case of reversed polarity. For the rest, it is plain that gravity is not of great importance in the mechanics of development.

In centrifugal force we have another force which, like gravity, acts in such a way as to "arrange" mixtures of lighter and heavier substances, the heavier parts being separated centrifugally from the centripetally-placed lighter ones. Thus, by regulating at will the strength of the centrifugal force, we can separate the substances of the egg from one another more sharply than is possible by gravity. It is found, in fact, that when the egg of the frog has been subjected to strong centrifugal force, the yolk-concentration at one pole of the egg increases to such an extent that it opposes an insuperable obstacle to cleavage. The egg, as a result, undergoes partial cleavage, the pole deprived of yolk being divided into cells, while the abnormally yolky one is not divided at all. The manner of cleavage of the centrifuged frog's egg in this way comes to resemble that of such eggs as those of Birds, where the cleavage

PLATE VII

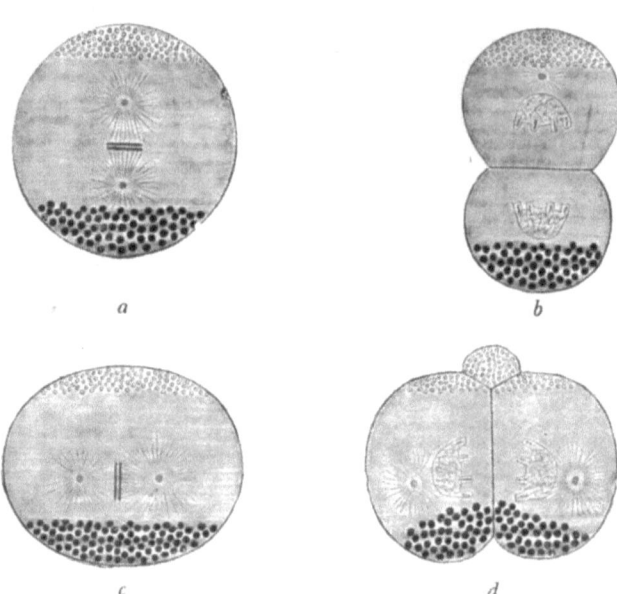

FIG. 94, *a–d.*—First cleavage of the egg of *Ascaris*: the normal process (*a* and *b*), and the formation of a "ball-egg" under the influence of centrifugal force (*c* and *d*). (From Schleip, 1929, after Boveri.)

is normally discoidal, the yolky region remaining unsegmented because the mass of yolk overcomes the forces of cleavage.

By making it possible in this way to influence the course of cleavage, centrifugal force has become an important instrument of experimental research; by its aid it is possible to alter both the direction and the rate of cleavage, a procedure of importance in studying the potency of the blastomeres which it affects. It may be added that centrifugal force can change the shape of the egg, and since the cleavage-spindle generally lies in the direction of the greatest protoplasmic mass, the direction of cleavage is also, in certain circumstances, involved.

The ripe egg of *Ascaris megalocephala* exhibits a more or less marked polar differentiation, manifested by a symmetrical arrangement of the cytoplasmic substances and inclusions about the primary egg-axis. In normal cleavage the first division yields two unequal blastomeres (Pl. VII, Fig. 94, *a, b*), the cleavage-plane being at right angles to the primary axis of the egg, and parallel to a hypothetical stratification. The dissimilarity of the two blastomeres is made evident by their fate: one is the first somatic cell, the other is the first "stem-cell" (cf. above p. 72). When strongly centrifuged, the substances of the egg are more completely separated; and since the egg, on account of the relative densities of its materials, comes to lie with its primary axis in the direction of the centrifugal force, there occurs at the same time a flattening of the egg in the direction of this axis. If, now, the egg begins to cleave while in the centrifuge, the cleavage-spindle places itself permanently in the direction of the greatest diameter of the egg. As a result, the cleavage-plane passes along the primary egg-axis—i.e., is at right angles to the normal cleavage-plane (Fig. 94, *c*). Thus, the two blastomeres formed by cleavage in the centrifuge are similar, for each has a portion both of the animal and of the vegetative cytoplasm of the egg. When, by rapid centrifuging, the granular inclusions in the cytoplasm are all crowded together at the animal pole, the cleavage-spindle is shifted bodily towards the vegetative pole (Fig. 94, *c*); its sphere of action no longer reaches effectively as far as the animal pole; and the inert masses concentrated there form an obstacle. The cleavage-plane therefore does not pass straight through to the animal pole, but is deflected on either side by the granules. The result of this is the production not only of the two blastomeres, but also, at the animal pole, of a "ball" consisting principally of cytoplasm filled with granules (Fig. 94, *d*). Further study of the effects of modifying cleavage in this way by means of the centrifuge may be expected to yield interesting results, though no specific action of centrifugal force appears to exist.

(b) *Temperature*

Temperature, again, is a factor of the normal environment which plays an important part in development—as, indeed, it does in all the processes of life. This is shewn by the presence of an upper and a lower limit of temperature, to transgress either of which causes the cessation of the processes. Between this maximum and minimum, the values of which vary with the kind of organism, lies the optimum, or most favourable temperature. In general, a lower temperature retards development or interrupts it altogether, a higher temperature accelerates it; and if the temperature departs too far from the optimum, development is inhibited. In connexion with this we may use the expressions heat-rigor and cold-rigor.

Temperatures departing from that which is normal produce numerous special reactions, leading to disturbances of morphogenesis. If, for example, sea-urchin eggs, ten minutes after fertilization, are subjected to a temperature of from 30° to 31° C., the germ as a whole is disintegrated: the individual blastomeres tend to separate from one another. As the result of complete separation of the two ½-blastomeres, twin-formations may arise. Gastrulation also is easily disturbed by temperatures higher than the optimum, the invagination of the archenteron being either delayed or completely prevented. The latter condition, which can be produced in the sea-urchin egg by employing a temperature of 30° C., gives rise to a so-called exo-gastrula. In this the archenteron, instead of being tucked in, is evaginated as an appendage, which in the end may completely degenerate. Since the other processes of development remain unaffected, larvæ may thus arise which are without a gut. This and other disturbances of gastrulation may possibly be attributable to changes in the viscosity of the cells, due to variations in temperature. In the Amphibia there generally result, later on, all kinds of monstrosities; and these have also been observed in the case of the hen's egg as a consequence of going beyond the temperature-maximum.

Temperature acts also upon the embryo *sensu stricto*. Its importance in relation to changes of form in many insects has been closely investigated, especially in butterflies. These often shew the phenomenon of seasonal dimorphism. By this is understood the appearance, during the course of the year, of several different forms, distinguishable from one another principally by their colouring, and each one characteristic of a definite season. At the same time it is observable that quite similar variations are connected with a warmer or a colder climate, so that there is a definite geographical distribution of these varieties. It seemed obvious, *a priori*, that a dimorphism of this sort must be connected with the temperature differences of the seasons and of the different geographical regions. Experiment confirms the truth of this assumption.

To cite a few examples: by the action of heat on the pupæ of *Vanessa urticæ* the colour of the butterfly can be changed to that of the southern variety, found in Sardinia and Corsica, while the action of cold changes it to the Lappish colour-variety. In the "warm" variety the red ground-colour becomes more glowing and darker: the dark elements of the pattern are reduced. Cold makes the ground-colour of the wings lighter, and distributes the black scales more widely. Frost and heat alike produce a conspicuous darkening. In *Vanessa antiopa* the action of heat is to broaden the yellow marginal band, to make the blue spots smaller, and to enhance the striping of the forewings, as in the variety of this butterfly found in Mexico and Guatemala. Fig. 96, Pl. VIII, shews the action of heat on this butterfly in comparison with its average condition (Fig. 95).

After heating the pupæ longer, the yellow marginal band becomes darkly speckled. Cold makes the ground-colour of the wings lighter; in certain cases it becomes yellowish-red instead of dark brownish-red. The "warm" variety of the same butterfly has its yellow marginal band very wide and, occasionally, speckled with black.

Something very similar can be effected in the case of butterflies which exhibit a seasonal dimorphism. By the influence of heat on the hibernating pupa, the summer form is produced instead of the spring form; and, conversely, after the action of cold on the summer pupa, the spring form instead of the summer form emerges.

Since the temperature of the environment certainly influences metabolism, growth is also affected by it, not only in poikilothermal but in homoiothermal animals.

In warm surroundings (33° C.), for example, the external gills of frog larvæ become larger and more branched than in the cold (15° C.). In the case of animals in the warmer and colder cultures the pronephros is bigger than it is in normal cultures (23° C.), though not as a result of an identical reaction to heat and cold. In warmth the larger size is due to the great development of the kidney tubules, in cold to the appearance of lymphoid tissue. The cooled animals have a larger liver than the warmed ones, but this again is not a simple difference in size, for in the cooled animals the vascular system of the liver is much less developed than in the warmed, though development of the trabeculæ is greater. Thus the unequal development of particular organs is due to unequal differentiation. Taking into account altered functional requirements in connexion with metabolism, an explanation must be looked for in the specific mode of reaction of individual tissues to different temperatures.

In young albino rats the length of the tail compared with that of the body increases with rise of the external temperature. If the animals, when from three to four weeks old, are placed in a cooler environment, there is a relative shortening of the tail during the next nine to eleven days, due to the greater growth of the trunk as compared with that of the tail.

Temperature, in these cases, acts indirectly through changes in the temperature of the body, which rises and falls with that of the environment. That the relative length of the tail is thus correlated with the body-temperature appears from the fact that reduction of the body-temperature by means of a febrifuge (e.g. antipyrin) produces the corresponding effect. The relative change in the length of the tail is plainly a result of the altered rate of metabolism of the body as a whole. This is confirmed by results of the investigation of the greatly changed internal organs of animals kept at very high temperatures.

(c) *Light*

Among the environmental factors which are everywhere present light is one that plays a part in development. Though a definite amount of light is not, indeed, a *sine qua non* for the development of the animal germ as in the case of heat, the germ is very sensitive to irradiation with light of short wavelength, and visible rays often have a marked effect upon it.

Ultra-violet light, for example, inhibits cell-division; indeed its action causes retrogression of a division that has begun. There is thus a light-rigor of development, just as there are heat- and cold-rigors. Visible rays can inhibit in the same way, though to a lesser extent. Intensive irradiation with ultra-violet light always causes great damage to embryonic cells; and after prolonged irradiation they die. This is of great technical importance in experimental embryology.

The effect of light is especially obvious in the formation of pigment. In insects particularly, pigment reacts in a striking way to the different wave-lengths of light which to the eye appear as different colours.

This is beautifully shewn by the pupæ of the Cabbage White, *Pieris brassicæ*, and by other butterflies. The coloration and pattern of the *Pieris* pupa are due, in the main, to the presence of a white and a brownish-black pigment. The former has its seat in the cells of the hypodermis, where it is evenly distributed in the form of small granules; it forms the white ground-colour of the pupa. The black pigment is secreted in the outermost layers of the chitinous cuticle to form small patches, dots, and streaks. These make a definite design. The general appearance of the pupa when seen from a little distance is grey; the integument is opaque, and the deeper tissues cannot be seen through it. There are, in addition, small masses of yellow and of blue pigment, but these do not really affect the colour-scheme.

The formation of the white and black pigments depends, to a very great extent, on the kind of light to which the caterpillar is exposed shortly before the onset of pupation. Depending on the character of this light, different modifications of the colour-scheme arise: one kind of light favours the formation of white and (particularly) of black pigment, while light of other wave-lengths inhibits their secretion. This inhibition is at times so strong that the white pigment only tinges the ventral side of the pupa, while on the dorsal side it is only formed in a few definite places. At the same time the black pigment of the pattern, in both the number and the size of its spots, is greatly reduced. The lack of white ground-pigment makes the integu-ment transparent, so that the colour of the deeper tissues and of the hæmolymph can be seen. This, in normally pigmented pupæ, is greenish, but in pupæ with reduced pigment it is of a specially intense green, so that these pupæ appear bright green, and sometimes even dark green.

To produce these several modifications of pupal colouring it is only necessary to allow the caterpillars to pupate on back-grounds of different colours. The pupæ on a given background-colour are not, indeed, completely uniform, but the over-whelming majority of them belong to a definite type as regards colour and pattern, and the remainder tend on the whole towards this same modification. Fig. 97, Pl. IX, exhibits several colour-modifications of the *Pieris* pupa; the differences due to variations in the development of the black pigment stand out with special clearness; the green coloration of the pupæ with reduced pigment does not appear in the black-and-white reproduction. On black, dark grey, and red backgrounds there arise—for the most part—pupæ with much black pig-ment, and with an opaque white ground-colour (Fig. 97, *a, b*).

The darkest form is rare, and is produced generally on a black background—sometimes also on a red or a grey one, though the characteristic variation on the last two colours is that shewn in the second illustration (Fig. 97, *b*). Blue and yellow environments give rise to pupæ less strongly pigmented (Fig. 97, *c*); the ground-coloration is, however, still preponderatingly white, sometimes with a slight greenish appearance on the dorsal side. On a green background there occurs, together with the greater reduction of black pigment, an extensive reduction of white on the dorsal side and on the flanks of the pupa (Fig. 97, *d*), so that this modification, all things considered, must be classified among the green pupæ. The greatest reduction of pigment is brought about by an orange-coloured background (Fig. 97, *e*); the reduction of the dark elements of the pattern can go even further than in the pupa illustrated; the white pigment has almost completely vanished, hence the pupæ are bright green, or in some cases quite a dark green.

Some indication of which constituents of the light-environment are active is gained by a comparison of these modifications with those of pupæ reared, not simply in a coloured environment with access at the same time of diffused daylight, but under coloured irradiation, with daylight excluded. This is most simply done by culturing caterpillars under suitable light-filters. Blue light, in this case, has the same effect as a blue background, an orange light as an orange background; but red light, unlike a red background, produces pupæ which are mainly green, though not to the same extent as does orange light. Darkness, i.e. complete exclusion of light, does not produce a complete suppression of pigment, as might be expected, but a definite inhibition of the black pigment.

Since red background and red irradiation have different effects, the active part of the light is that which is absent in the latter conditions but present in the former; this is the ultra-violet constituent of daylight, which cannot pass through a red light-filter, but is reflected from a red background just as it is from a black or a grey. The infra-red rays may also be of importance. Ultra-violet is concerned primarily with the black pigment, infra-red with the white. Removal of the ultra-violet causes the suppression of the black pigment, weakening of the infra-red results in a lessening of the white pigment. Increasing the content of ultra-violet in the light-environment increases the black pigmentation. Now we must not conclude from this that ultra-violet and infra-red are alone decisive; for then darkness would be expected to involve complete de-pigmentation which, as we have seen, is not the case. It is much more likely that the general quality of the light used, characterized by a specific mixture of wave-lengths composing it, is what exerts the particular action on pupal pigmentation, and that the ultra-violet plays a special part in connexion with the black.

The very interesting fact has been established that the different colour-types of pupæ can be distinguished from one another by the differing chemical composition of their hæmo-lymphs and of their tissues. The action of light has thus a profound effect upon the whole character of the pupa.

The pupæ of other butterflies also react to the light-conditions of their surroundings in a comparable way; for example, the Vanessidæ form golden-coloured pupæ when on a bright gold surface.

It is important to notice that in order to produce the above reactions it is necessary for the light-factor to be active only during a comparatively short, but definite, period in the life of the caterpillar; and further, that the eye of the caterpillar plays an indispensable part in the appearance of the reaction. Only if the coloured light falls upon the eyes of the caterpillar does it affect the formation of pigment in the pupal integument. The sensitive period begins at the moment when the caterpillars cease to eat and prepare for the pupal ecdysis. It therefore makes no difference whether the caterpillars are kept in the coloured light during their whole life, till after pupation, or whether they are kept in that light only during its period of activity. In what manner the eye—and presumably also the nervous system—comes into the series of reactions is not really clear. The butterflies which emerge from pupæ thus modified by the

PLATE VIII

FIG. 95.—*Vanessa antiopa*, normal.

FIG. 96.—*Vanessa antiopa*, modified by the action of heat on the pupa (periodic warming to 44 C.).

PLATE IX

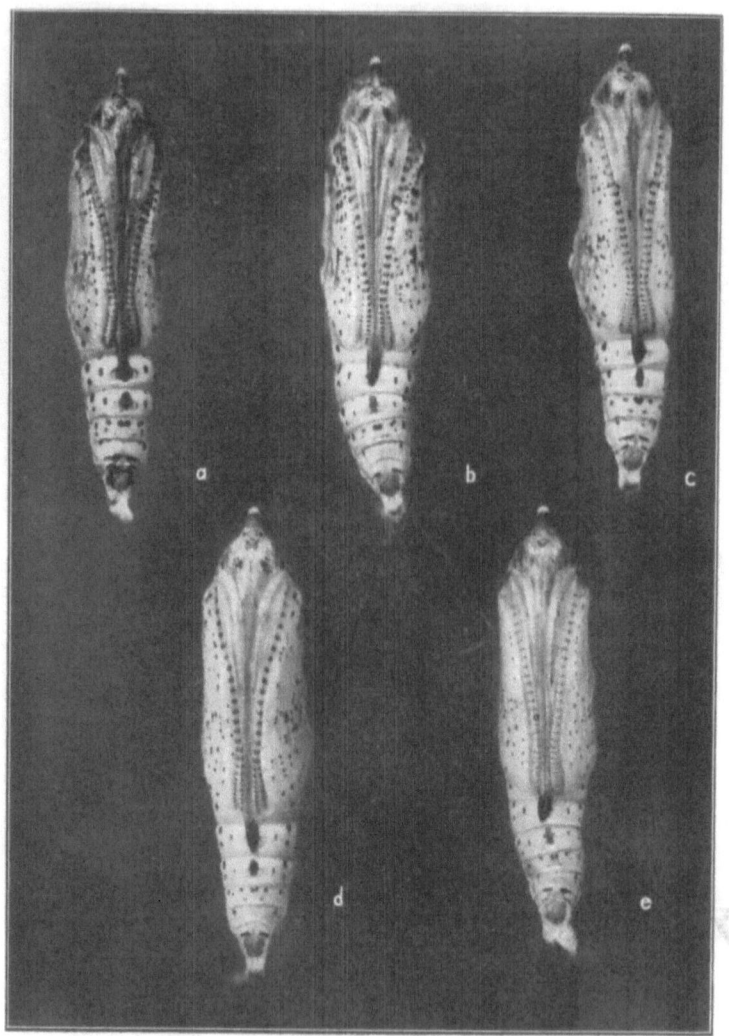

Fig. 97, *a–e.*—Five modifications in the pupa of *Pieris brassicæ*. Living animals photographed from the ventral side. *a*, form with most pigment, resulting from pupation on a black—occasionally also a red or dark grey —background; *b*, commonest modification on a red or grey background; *c*, commonest modification on blue or yellow; *d*, on green; *e*, on orange. The ground-colour of forms *a–c* is white, that of form *d* mainly green, that of form *e* completely green.

action of light are not themselves affected in their pigmentation.

The coloration of other insects is affected by the light of the environment. This is true of the Stick-insect, *Dixippus morosus*, which, on the whole, reacts like the *Pieris* pupa. Numerous experiments have also been carried out in other groups of animals, including the Vertebrates. In the most diverse forms, the action of light can cause the development of pigment in normally pigment-free individuals or regions. Researches on the salamander have led to much discussion. The colour-scheme of this animal, as is well known, is one of black and yellow patches. The action of yellow light is to increase the yellow pigment, while a black background causes an increase in the black pigment. The results of different workers are to some extent contradictory, and no conclusion can at present be stated.

In certain circumstances light has an influence not only on the colouring of animals, but upon their structure and growth. Hydroid colonies often shew a pronounced heliotropism, which determines their direction of growth; while in certain forms (e.g. *Eudendrium racemosum*) it is only in the presence of light that hydranths are regenerated.

Among vertebrates an interesting case is known in which the formation of the eye can be brought about by the influence of light. *Proteus anguineus*, a cave-dwelling animal, has extremely rudimentary eyes, which are covered by the thick integument of the head. If very young animals are exposed to the light for several years, the eye becomes fully formed in all its parts. Conversely, in the larva of the Axolotl complete absence of light definitely inhibits the finer differentiation of the retina. We see in these examples, if not a direct effect of light, at least its participation in the normal formation of an organ.

Since processes of development are extremely sensitive to rays of short wave-length, it is probable *a priori* that they will react to wave-lengths still shorter than the ultra-violet, such as X-rays and radium rays. Though in this case we are no longer dealing with factors of the normal environment, the matter is interesting from the point of view of method and technique.

In an earlier section (p. 50) we dealt with the damage caused to the nucleus of the reproductive cells by the rays in question; and it was there explained how cells enucleated in this way could be used to produce artificial parthenogenesis. If the irradiation goes beyond a certain point, there appear, in later development, extreme malformations, which are to be regarded as the results or concomitants of injury to the nuclear substance. Frequently cells are eliminated; the closure of the blastopore is impeded, so that a giant yolk-plug persists; the development of the neural tube is adversely affected, and so forth. The injury is especially marked in the case of the eyes. After irradiation of the reproductive cells with radium bromide, or with "hard" X-rays filtered through copper and aluminium, there is sometimes found in the embryo, instead of an eye, only a small strand of pigmented, partly degenerated cells. Since X-rays are of so much importance in the practice of medicine, such results are of direct interest in that connexion.

Later stages of development are also extremely sensitive to such rays. Though development at first proceeds normally, great disturbances appear later. In this case it is the central nervous system, the sense-organs—and, again, the eyes in particular—that are principally affected. Next to these comes the heart, while the auditory vesicles, the muscles, kidneys, and notochord are more resistant. Here, too, the injuries are due to damage done to the nuclei of the cells.

(d) Chemical composition of the medium

So far we have been concerned with the physical components of the environment; it is obvious that the chemical composition of the medium in which development proceeds cannot be without importance. This holds especially for the germs of those animals that lay their eggs freely in the water. For here the medium is subject to greater and more complex fluctuations than it is in the case of eggs which develop on dry land, surrounded only by the atmosphere. The environment of germs which develop within the maternal body is also complicated in its chemical composition; but the conditions of the environment, from their very nature, remain fairly constant. These germs are not, however, so readily accessible for experimental study as those first mentioned. Our interest is therefore chiefly centred in those eggs which develop in water.

The chemical composition of the environment is, of course, most complicated in the case of marine animals; and it is found that any considerable change in the composition of the sea-water influences development. In particular the absence of certain definite substances leads to disturbances. To give one example only, absence of calcium salts affects the egg of the sea-urchin by causing the blastomeres to separate from each other, though cleavage is otherwise normal. The result is similar when there are no sodium and potassium salts present. If such an artificial sea-water be allowed to act only up to the two-cell stage, and the egg be then put back into normal sea-water, the two separated blastomeres develop independently, and twins are formed. The cells can also be separated from one another in the same way in later stages of cleavage and development. Probably the reaction we are witnessing here is one involving the colloid chemistry of the surface of protoplasm.

An excess of certain salts also disturbs development. For example, if the composition of the sea-water is changed by the addition of a small amount of a 3·7 per cent. solution of lithium chloride, the archenteron of the sea-urchin germ, instead of being invaginated, is evaginated. There arises in this way an exo-gastrula, such as we have mentioned in connexion with the effect of temperature. The germ assumes a dumb-bell shape—a roughly spherical ectodermal part being separated by a deep constriction from a similarly-shaped endodermal part. The endodermal part may increase in size at the expense of the ectodermal, so that the latter is completely converted into endoderm. After what has been previously said about the potency of the primordial rudiments, this conversion will not be thought incomprehensible. The exo-gastrula does not owe its formation to the specific action of lithium: it can also be produced by the action of other salts. Similar phenomena can occur in eggs which develop in fresh water, such as those of the Amphibia. By the addition of small amounts (0·3–1·0 per cent.) of the chlorides of sodium, magnesium, or lithium the invagination of the vegetative region of the germ is delayed, and a giant yolk-plug is often formed; so that here again we have a kind of exo-gastrula. In extreme cases gastrulation is not completed. During later development it is especially the central nervous system which is involved; either the closure of the neural tube is prevented in that part of it which gives rise to the brain, or there appears, in the region of the spinal cord, a cleft due to the forcing apart of the neural folds by the yolk-plug (Fig. 98), that is, a *spina bifida*. In other cases the neural

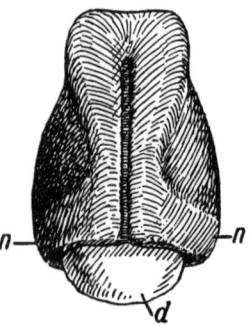

FIG. 98.—Neurula of *Rana esculenta* shewing *spina bifida* (splitting of spinal cord) as the result of treatment of the germ with a 0·7 per cent. solution of NaCl: *d*, persistent yolk-plug; *n*, neural folds.

tube is completely closed, and a tail-bud established, but the larva remains without a gut, because the endodermal region has not been invaginated. If the germ is replaced at the right time in normal water the less strongly marked deformations are regulated.

The sensitiveness of the rudiment of the central nervous system to changes in the salt-content of the surrounding water is also shewn by the appearance of *cyclopia* when magnesium chloride is added; though again we are not dealing with a specific action of that substance. In fishes, for example, when $MgCl_2$ is added to the water, the two primary eye-rudiments are drawn together in the median line, and there is actually formed only a single eye, set in the middle line of the head (cyclopia). After what has been said about the dependant nature of lens-formation, it will readily be understood that this large, unpaired eye possesses only one lens. Small doses of alcohol, chloroform, or ether have the same effect as MgCl$_2$. Among other deformities, *spina bifida* is frequently observed side by side with cyclopia.

The salt-content of the water can so influence morphogenesis

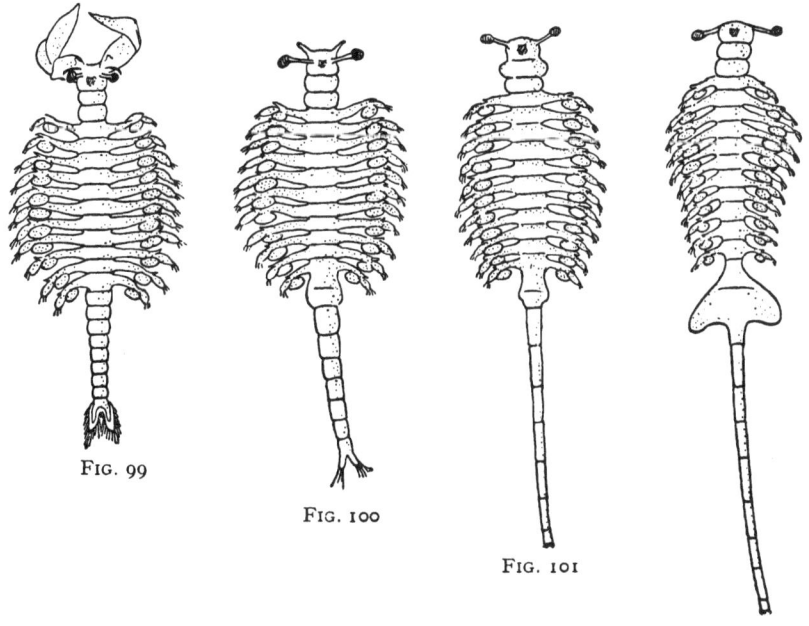

FIG. 99

FIG. 100

FIG. 101

FIG. 102

FIGS. 99 to 102.—Modifications of *Artemia salina*. Fig. 99, male from water with very low salt-content; Figs. 100 to 102, females from water with in-creasing salt-content—sp.g. 1·012, 1·090, 1·142, respectively. (After Abonyi.)

that the modifications which arise vary with its concentration. Figures 99 to 102 shew the different forms assumed by *Artemia salina* in relation to the amount of salt in their culture water. The more salt the water contains, the more slender the general appearance of the animal. The action is, however, only relative, in the sense that each individual does not react in the same degree to changes in the salt-content. The causal connexions involved in this reaction are not understood.

(e) *Nutrition and metabolism*

Food must be reckoned a factor of the environment, even though in general it can only become effective indirectly through the introduction of its constituents into metabolism. We are not concerned in this place to discuss the phenomenon that growth in general is influenced by the nature of the food, but the question as to whether special characters of the organism are directly connected with nutrition. The observations available are not numerous, and a few general indications must here suffice.

The pigmentation of butterflies can be altered in some cases by changing the food of the caterpillars. If caterpillars of *Lasiocampa quercus* are fed on sainfoin instead of on their normal food (oak-leaves) the coloration of the butterflies is markedly abnormal. The wings of the female become considerably darker; those of the male, very much lighter at the edges. This action, however, can hardly be a specific one. It is much more likely to be connected with the fact that the time of moulting and the period of the pupal state are postponed by the abnormal food; so that the length of time available for the production of wing-pigment is also changed.

Very active specific substances can be introduced into metabolism with the food. By feeding animals on the endocrine glands, the action of these glands on development can be studied. If the larvæ of Amphibia are fed on thyroid gland, growth is inhibited, but not development; feeding with thymus has the opposite effect—inhibition of development but not of growth. Metamorphosis is hastened by feeding with thyroid gland—an effect produced also by pure iodine. Regeneration of the tail in the tadpole is most strongly marked after thymus feeding, and is weakest after administration of thyroid. These things are principally of importance from the point of view of the technique of research, since the substances administered do not form a part of the normal food. The special character of the ordinary food of an organism can, however, affect the form which that organism assumes.

A clear demonstration of this is furnished by the effect of

nutrition on the length of the gut of the frog larva (Fig. 103, *a* and *b*). If the larvæ are fed on a meat diet only, the gut remains considerably shorter than it becomes when they are reared exclusively—or at least mainly—on plant food; this is evident from the number of coils of the gut. In the first case the cross-section of the gut is greater than in the second; but, in spite of that, the absorptive surface of the gut of meat-fed

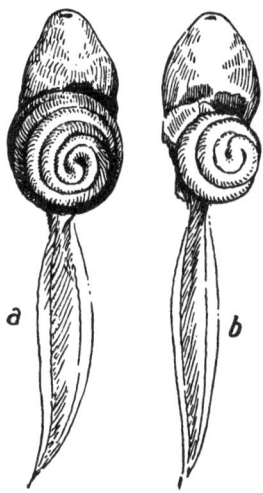

FIG. 103.—Two larvæ of *Rana fusca* (*R. temporaria*), one of which (*a*) was reared on a mixed diet (mainly vegetable), the other (*b*) on animal food only. The abdominal body-wall is removed to show the coils of the gut.

larvæ is only one-half of that of the plant-fed larvæ. This action upon the gut is not a direct or a specific one, but the consequence of altered function. Very abnormal food causes, in other animals too, a change in the length of the gut. As a result of increased functional requirements the surface of absorption is increased. In such cases we speak of *functional adaptation*. This phenomenon will often be encountered, if we do not neglect the functional aspect of the development of organs.

(*f*) *Functional requirements*

The form assumed by the spongy tissue of the long bones of man and other vertebrates bears unmistakably a functional relation to the compression and tension effects of the mechanical load. At the ends of these bones, where they form joints, their solid walls give place to a system of slender bony trabeculæ, the arrangement of which accords, in the main, with that of the lines of mechanical compression and tension. If in their embryonic rudiment these trabeculæ shewed no orderly arrangement, and this arrangement first appeared after the bone had begun to function, then obviously the action of the mechanical load could be held directly responsible for the appearance of the disposition in question. The arrangement of the spongy tissue in the fœtus, however, is already substantially the same as it is in the adult bone. Use cannot therefore be the only determining cause. That it is contributory, however, appears from cases where the architecture of the bone changes with changing demands made upon it; as, for example, after a crookedly-healed fracture, or in limb-bones abnormally placed as a result of anchylosis, in which the joint disappears by the fusion of the articulating bones. What is most interesting here is the fact that the transformation of the spongy structure occurs not only in the immediate neighbourhood of the fracture, or of the original joint, but extends to parts of the bone far removed from these places. The arrangement of the reconstructed architecture corresponds very well to the changed compression and tension curves. The manner in which functional activity is here effective is not yet clearly understood. The normal formation of the spongy tissue depends equally, then, upon its fitting harmoniously into the form of the bone as a whole (a thing which was already important in the embryo), and upon functional adaptation.

The form of a joint is also subject to functional activity. The form of the parts of the joint is certainly inherited as such, but it is clear, as the result of experiment and from surgical observations, that function too exercises a moulding influence. As an example of this we may cite a case in which the elbow-joint of a thirteen-year-old girl had to be opened up on account of anchylosis—the humerus having completely fused with the radius and ulna (Pl. X, Fig. 104, a). After the operation the

PLATE X

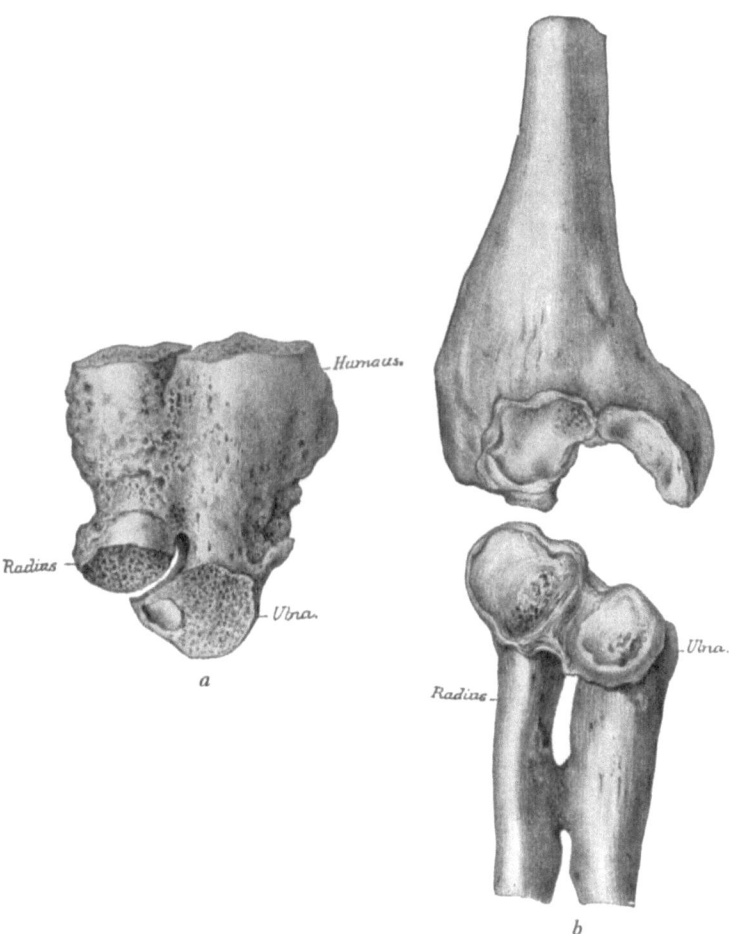

FIG. 104.—Formation of a secondary elbow-joint after resection of the original joint, which had become anchylosed. *a*, the fused ends of the bones removed by operation (humerus, radius, and ulna); *b*, the secondary articulating surfaces $2\frac{1}{2}$ years after the operation. (From Korschelt, 1927, after Czerny.)

arm could again be used. Since the patient died $2\frac{1}{2}$ years after the operation, an exact description of the condition of the joint was possible (Fig. 104, b). On the cut-surfaces of the three bones (humerus, radius, ulna) caused by the operation, new surfaces of articulation had formed, and the shapes of the ends of the bones approximated to those which normally characterize the joint.

The effect of a functional demand appears, again, when a piece of a vein is implanted into an artery, and is thus subjected to greater pressure; this produces a corresponding strengthening of its wall.

We will not multiply examples. Considering all that has been said, we may safely assert that the environment as a whole, and also its individual factors, intervene in the realization of development. An important question, which still remains to be discussed, is whether these actions of external factors are always confined to the individuals directly concerned, or whether they may not sometimes reach beyond the individual and affect succeeding generations.

II. THE TRANS-INDIVIDUAL ACTION OF EXTERNAL FACTORS

1. The Reaction-Basis, the Reaction-Norm, and the Course of the Reaction

In considering whether the action of external factors can extend beyond the individual immediately affected, we must disabuse ourselves of the notion that this is simply the question of the inheritance of acquired characters in the Lamarckian sense. Admittedly the problem has generally been considered in connexion with discussions on volution, and especially in connexion with attempts at a Lamarckian solution of the problem of descent. Though no completely satisfactory solution on these lines has been found, this preliminary discussion has considerably narrowed down the problem. The question of the trans-individual action of external factors is a much wider one; and the Lamarckian problem of the inheritance of acquired characters—to use the current, though not very exact expression—is at most one special case in a whole series of problems. Moreover, the problem that primarily interests us here is one purely of experimental embryology. Whether, with this limitation, the results obtained will be found to have any bearing on the theory of descent, and in particular on the Lamarckian hypothesis, is a question in itself. But even if, from this last point of view, results should prove to be negative, the problem would remain an important one for analytical embryology. This is obvious from the fact that the experimental embryologist does not limit himself to apparently adaptive or "purposeful" modifications of the individual, or to such modifications as might possibly form a step in evolution, and consider whether or no these make themselves felt beyond the individual which is actually modified. On the contrary, he interests himself also in certain effects of external factors which are recognized as being definitely non-adaptive and transient.

A trans-individual action of the external factors certainly exists, and already several forms of this are known. In order to present our ideas clearly in this discussion, it is essential to start from quite definite concepts. We must take first the concepts of the reaction-basis, the reaction-norm, and the course of the reaction, and then concepts corresponding to the various possible kinds of induction.

Every organism receives from its ancestors a certain constitution, in the form of internal factors of development, which in some way are passed on as its hereditary complex. It is the interplay of these inherited factors with the external factors which produces the result in development. We may say that the separate internal factors *react* with one another, and with the external factors, during development—without intending to imply by this that the "reactions" are purely chemical. It is assumed, of course, that the same factor under the same conditions will always react in the same way; and this naturally applies to the whole of the developmental factors. Or, it may be expressed thus: the developmental factors (the hereditary complex) possess an absolutely definite *reaction-norm*. This is in itself unchangeable. If this were not so, not only would chaos reign in the relation between the generations, and the concept of heredity collapse, but we should lack the postulate necessary for every natural science. For the constancy of the reaction-norm of one particular natural factor is the assumption from which every investigation must always proceed. The trans-individual action of external factors can therefore never be based on a change in the reaction-norm.

What is first modified by change in the external factors is the *course of the reaction*; and this produces a result in development that varies according to which of the external factors interact with the internal factors and their definite reaction-norm. Such an alteration of the course of the reaction (and of its result) obviously possesses, from the point of view of development, an importance for the individual alone; and it is manifestly impossible for a changed course of reaction to alter in any way the reaction-norm.

Nevertheless a trans-individual action of external factors is possible through changes induced in the reaction-basis. This conception, which has already often been made use of in preceding chapters, denotes by reaction-basis that which is itself reacting, that is to say the actual seat of the internal factors of development. The various regions of the reproductive cell, considered as constituents of the reaction-basis, must certainly be both quantitatively and qualitatively unlike; but it would be wrong to regard particular pre-designed constituents of the nucleus as actually constituting the reaction-basis. The reaction-basis consists, rather, of all those constituent parts of the cell which are really specific: the whole cell, apart from its non-specific inclusions, is the reaction-basis. This reaction-basis cannot be regarded as fundamentally unalterable or, in its several parts, inviolable; and a belief in the theory of descent involves the tacit assumption of this mutability.

Alteration of the reaction-basis naturally carries with it, as a secondary consequence, a change of the reaction-norm. It is not, however, that the original reaction-norm has itself been modified, but that there is a new reaction-basis. For there exists in connection with each specific reaction-basis a specific reaction-norm.

That a change in the reaction-basis should occur through modification of the course of the reaction is, to say the least, improbable. It is more possible that such change is brought about by influences which affect the reaction-basis directly, or by a modification of the end-product of development—for the reaction-basis of the next generation would thus find itself in a new situation. This new reaction-basis is not simply contained within the end-product of development, but forms an essential part of it; so that the further possibility arises that the conditions of its new situation may affect the reaction-basis.

2. The Various Possible Forms of Induction

What is meant will become clearer after we have discussed the different possibilities of induction. If we take induction to mean the bringing about of a definite condition by a factor working from without, then, theoretically, it is possible to imagine seven different forms of the process (Fig. 105, A–G). It must be explicitly stated, however, that all the possibilities are not in this way exhausted; nor, on the other hand, are all these possibilities realized in nature. At the same time, the experiments so far performed can be explained by bearing in mind the possibilities here mentioned.

The simplest form is *direct induction*. The relations are somewhat different in the Protozoa from those in the Metazoa. In the Protozoa the external factor can directly affect the reaction-basis itself, and in certain cases transform it (A, a); for in this case the organism itself and the reaction-basis, which must be regarded as the foundation of the succeeding generation, are to all intents and purposes identical. In the Metazoa the external factor, in order to influence the reaction-basis of the succeeding generation contained in the soma, must work *via* the soma itself. Inasmuch as there are factors which can pass unaltered through the tissues lying between environment and reaction-basis, a direct change in the reaction-basis may

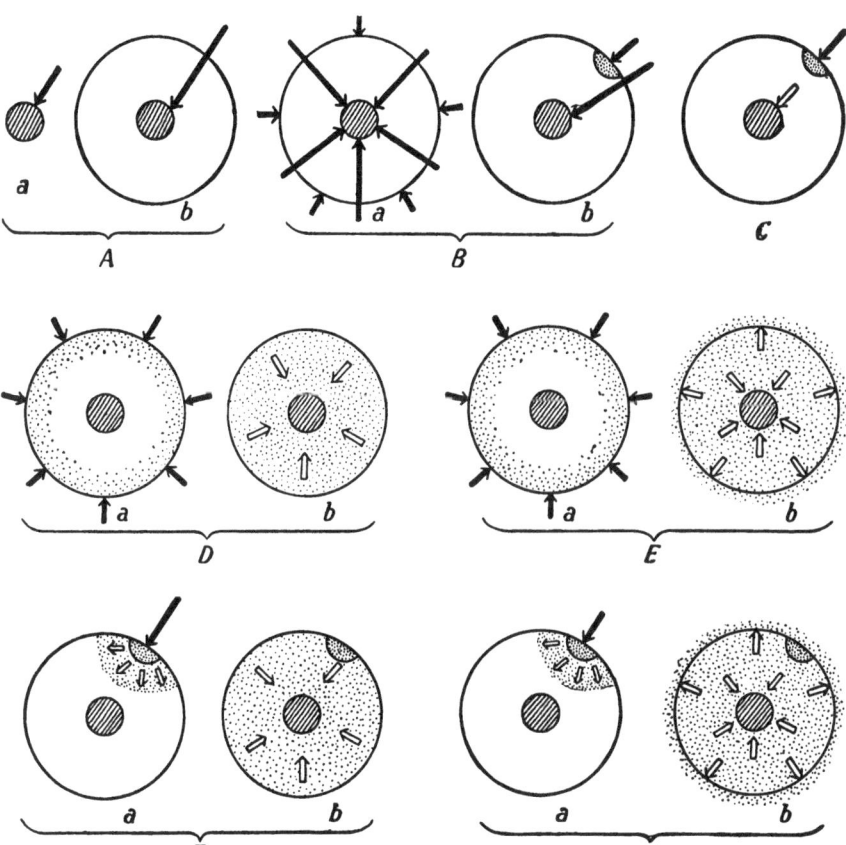

FIG. 105, *A–G.*—Diagrams to explain the various possible kinds of inductive action of external factors. *A*, direct induction in the Protozoa (*a*) and Metazoa (*b*). *B*, parallel induction; *a*, general influence on the whole soma and at the same time direct inductive action on the reaction-basis of the following generation; *b*, special influence on a part of the soma with simultaneous direct inductive action on the reaction-basis. *C*, Merogenic somatic induction (true Lamarckian induction). *D*, simple hologenic somatic induction, without the changes induced in the soma reacting on the visible condition of the soma; *a*, first phase: alteration in the general nature of the soma; *b*, second phase: action of the changed soma on the reaction-basis contained in it. *E*, *a*, *b*, hologenic somatic induction, but with reaction on the visible condition of the soma—this reaction is indicated by stippling. *F*, specialized hologenic somatic induction (apparent merogenic induction), *a*, first phase: action of the external factor on a special part of the soma, from which an influence passes out to the whole soma; *b*, second phase: action of the altered condition of the soma on the reaction-basis. *G*, specialized hologenic somatic induction, but with reaction on the visible condition of the soma, *a*, first phase as in *F*,*a*; *b*, second phase: action of the altered condition of the soma on the reaction-basis contained in it, and at the same time on the visible condition of the soma itself. The outer circles indicate the soma, the inner, shaded circles, the reaction-basis.

come about in this way, and without the soma itself being thereby altered (*A*, *b*). Induction, except of this direct sort, seems to be impossible in the Protozoa; though the relations, in particular cases, are more complicated than can be shewn in a presentation which is necessarily diagrammatic.

Parallel induction is often discussed in the literature. What is meant by the expression is the influence of an external factor, at the same time and in the same sense, upon both soma and reaction-basis. Here we may distinguish two possibilities: first, the external factor may affect the soma as a whole, changing its nature as a whole (*B*, *a*); second, the modifying factor may seize upon some particular part of the soma—a particular organ or character of the body (*B*, *b*). However that may be, the alteration of the soma, or part of the soma, is without effect upon the reaction-basis, for this is changed in the same sense, and independently, by the same factor. How far such a notion may be supported by actual observation will not be discussed; we shall say no more than that the frequency and importance of parallel induction are often overestimated.

As against the forms of induction hitherto mentioned, in which we imagine the external factor to bear directly upon the reaction-basis, we must understand by somatogenic induction, or *somatic induction*, those kinds of action in which the soma intervenes actively between the modifying factor and the reaction-basis. This category, again, may be subdivided.

In the first subdivision we may place what is appropriately called *merogenic* somatic induction. We must imagine here that the external factor comes into contact with a certain part of the soma, and modifies it; from this modified part (hence "*merogenic*") there then goes out—as the second phase of the process —a corresponding influence to the reaction-basis; and this last is thereby changed in such a way that in later development it reproduces the same somatic modification, but in the absence now of the modifying factor (Fig. 105, *C*). We have here, therefore, the conception of the "inheritance of acquired characters"—in the special Lamarckian sense of the words. It is more than doubtful whether, in actual fact, the process of induction is ever like this; in any case, weighty objections can be urged against it on general grounds, and up to the present nothing of the sort has been experimentally established.

With merogenic somatic induction can be contrasted *hologenic* somatic induction, of which we can again distinguish various forms. In order to review these possibilities of action it is best to divide each induction-process into two phases, one following the other (at any rate, in theory); we need not here consider whether, in actual fact, this chronological separation of the two phases always exists. The simplest case of hologenic somatic induction is that in which in the first phase the modifying factor changes the soma in general (D, a) without the change becoming patent. The second phase of the process (D, b) consists in an influence being exerted by the changed character of the body as a whole (hence "*holo*-genic") upon the reaction-basis contained in it. Whether or no the change produced in this last is of the same kind as the modification of the soma need not be considered here; we are concerned only with the way in which the reaction-basis can be influenced.

In a second form of hologenic somatic induction (E, a, b) the process would be exactly similar, were it not for the fact that in this case the soma, on account of its stage of development, is capable of being influenced; the change in its whole character is consequently made manifest in its outward form, a definite modification being produced. In such a case, what is induced, in the reaction-basis, is not this last modification, however, but the whole change of the soma, of which the perceptible modification is but the expression. If, then, in the next generation, the same modification appears in the soma as a result of the altered reaction-basis, this is not because it has become hereditary in the Lamarckian sense, but because the same general condition of the soma must of necessity appear again; and it is this which then produces the modification as a secondary result.

Now the whole process may conceivably take place in a more complicated way as follows. In the first phase (F, a) the modifying factor comes into contact with, and modifies, only one definite part or characteristic of the soma. This modification of a part results in the alteration of the whole character of the soma. In the second phase (F, b), this alteration affects the reaction-basis so as to induce in it a corresponding change. In development—as a result of the presence of such a reaction-basis—the modified somatic condition arises again, and the outward expression of this is the formation of the original modification at one particular place. It is clear that in this way a merogenic somatic induction may be simulated, such as forms the basis of Lamarckianism; but it is equally clear that it is spurious, and that we are really dealing with a case of hologenic induction.

Another possible complication must be considered (G). In the first phase of the induction process a definite part of the soma is affected by the external factor concerned; whether a visible change is caused or not is unimportant. The affected part now exerts a general influence (G, a), which results in a change of the general character of the soma. This general change now acts (a) on the reaction-basis, inducing in it a corresponding change; and (b) it reacts on the soma, which is still capable of development. In doing this it influences not the part first affected but other parts, or characters, which now become definitely modified without the original modifying factor being immediately responsible (G, b). A trans-individual effect of such hologenic induction may come about in this way: first the new, developing soma exhibits the changed general character; then, as the result of the modified condition, the same kind of special modification again appears, secondarily, and at the same "site of reaction" where the original modification was manifested. In this case also there is apparently merogenic somatic induction—but it is apparent only. Not only does the action of induction not proceed from a modified character, but this character itself is the result of an alteration otherwise produced.

As we have said earlier, there exist further theoretical possibilities of induction, but these need not be considered here. There are, in the main, three chief variations of the process: direct induction—which possibly coincides with parallel induction—merogenic somatic induction, and hologenic somatic induction.

We at once see, in connexion with the above processes, that instead of the question as to whether an acquired character can be inherited, we are confronted with the other question as to whether a change of the reaction-basis is possible as the result of the action of external factors—either by direct induction, or with the soma introduced as intermediary. If this question is answered in the affirmative, it remains to be decided whether the change may be a temporary or a permanent one. If it is permanent, then a mutation has been induced, that is to say the change has been incorporated once and for all into the inheritance. If it is temporary, then the original modification has only been extended over several generations; it is in fact a persistent modification, which finally disappears from the reaction-basis and from the phenotypic "picture" of the adult organism. This does not depend on whether the modification or mutation produced stands, or does not stand, in a specific relation to the external factor originally active; nor does it depend on whether this persistent modification or mutation has, or has not, survival-value; but it does depend on whether such changes are induced directly or are somatogenic. The supplementary question may be asked: are there cases which are covered by the Lamarckian hypothesis, if not as regards their internal mechanism, at least when judged by the actual external effect? Or must we come to the conclusion that the statement of the problem of Lamarckism is still warranted only on the condition that its old conception of induction be given up, and that it adapt itself to the newer conceptions? These are questions which naturally interest the student of causal embryology less than they do the general biologist.

3. The Results of Induction

Leaving finer distinctions out of account, we may classify the results of the trans-individual action of external factors under three different heads: pre-induced modifications, persistent modifications, and mutations. Each of these phenomena can be conditioned by direct and by hologenic somatic induction.

(a) Pre-induced modification

We speak of a pre-induced modification if the change has been called forth by the action of an external factor upon the organism in its original unicellular condition, and if the modifying factor is no longer active during later development. The phenomenon is particularly striking if the pre-induction occurs while the egg is still within the body of the mother. There is nothing to indicate that a change in the reaction-basis is caused in this way; on the contrary, it is evident that only the course of development is modified. It is further characteristic of this kind of modification that it affects only the one generation immediately following, and then suddenly disappears.

We may mention, in this connexion, experiments on the form of the "helmet" of Daphnids, which is dependent on various external conditions (Fig. 106). *Daphnia longispina* possesses only a low helmet in its normal environment—the cold water of mountain lakes. If this species is reared in a hot-house, with more abundant food, the descendants are long-headed (high helmet): a modification has in fact arisen. If, after two years of this kind of culture, the animals are replaced in their original conditions of life, the first of the generations now formed is still long-headed; but with the next generation the long-headedness of the original race reappears. Conversely, the long-headed *Hyalodaphnia cucullata* can be made to produce short-headed descendants by the action of starvation and cold. When the latter animals are placed in favourable conditions the first generation produced is still short-headed, but there is no further after-effect of the modifying factor. The external factors have acted as favourable or unfavourable conditions, and have led to corresponding modifications in the generations directly affected. The nature of the first generation of descendants also constitutes a modification—produced by the same conditions. It was, indeed, induced in the germ-cells by the external factors; for the formation of these germ-cells took place under the favourable or unfavourable conditions prevailing at the moment. Naturally then, the developed organism reveals this favourable or unfavourable influence, and this

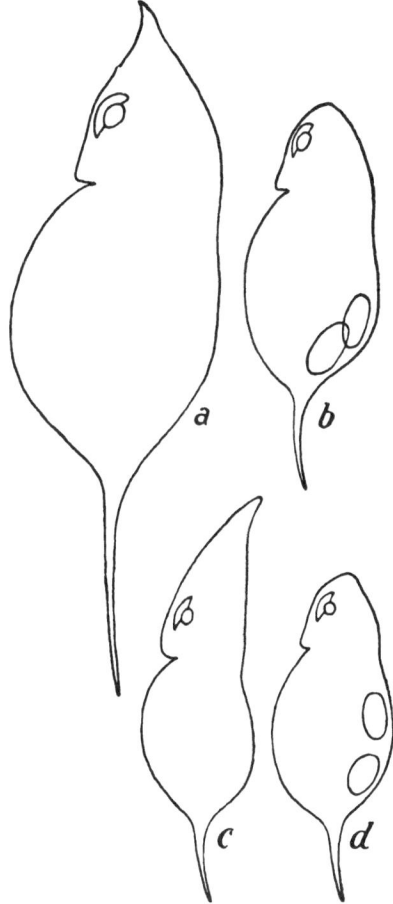

FIG. 106, a–d.—Long-headed and short-headed modifi-cations in *Daphnia longispina* (a, b), and *D. cucullata* (c, d). (After Woltereck.)

without there being even a temporary change in the real reaction-basis. The external factors with which we are concerned in these experiments (temperature, for example) can penetrate directly to the germ-cells; and since the condition of the egg is here clearly of prime importance, we have a case of direct pre-induction by the external factor. Some of the conditions of the environment may act in an indirect way through the soma (starvation or abundant food), so that in a soma which is stunted the eggs too are stunted, and similarly in the case of the opposite kind of reaction. The sudden cessation of the modification suggests that nowhere in this process is the reaction-basis involved.

(b) Persistent modification

In the case of true persistent modification the after-effect of the modifying factor does not suddenly cease, but there is seen a gradual diminution of the modified condition, and a gradual return to the original state when the modifying factor is no longer active. This must mean that the reaction-basis has here been influenced in such a way as to produce in it an unstable alteration. This can be caused by direct induction, or by the hologenic inductive action of the external factor. As examples of persistent modification we shall take only one case from the Protozoa and one from the Metazoa. Full justice cannot be done to the phenomena in the Protozoa owing to the brevity of our treatment.

There are numerous genetically distinct races of *Paramecium caudatum* and *P. aurelia*, amongst the distinguishing characters of which are varying powers of resistance to poisons and to extremes of temperature. The degree of resistance to poison (e.g. to arsenious acid) cannot be altered by selective breeding. It is possible, however, by gradual acclimatization to stronger doses of poison and to higher temperatures, to render these infusoria more resistant. In certain experiments arsenious acid was employed at concentrations $\frac{1}{3}$ to $\frac{1}{2}$ of that which was lethal for the *clone* (i.e. descendants of a single individual) which was used. Calcium compounds were used in the same way, and also a *Paramecium*-antiserum obtained from the blood of a rabbit after the injection of the animal with paramecia. After long-continued action of the solutions, and sometimes by a considerable increase of their concentration, *Paramecium* acquires a marked increase in its resistance to poison. But it is lost again in normal conditions in a poison-free medium. This does not occur immediately where there is only asexual reproduction; in that case the resistance persists through several generations, though it gradually dies away. The increase in resistance to calcium compounds is especially enduring; it even persists through several asexual divisions and through one conjugation, while the increased resistance to other poisons disappears suddenly with the first conjugation. Essentially the same results have been given by experiments of this kind on *Arcella*.

We have, without doubt, in the above instance a specific reaction to an external factor, and an adaptation. Accepting what has been previously said, the whole protozoan body must be regarded as the reaction-basis. From this foundation the new generation arises, and it is clear that this shows a change of character. This change, however, is not permanent; it is unstable, and generally disappears gradually, though sometimes suddenly (e.g. after conjugation). The modification originally acquired remains in existence, however, longer than the individuality of the infusoria which were actually influenced. Thus we have a persistent modification, directly induced by the external factor, since the latter acted directly on the reaction-basis.

There are also persistent modifications which are produced by hologenic somatic induction. As an example we may take the colour-varieties of the *Pieris* pupa, which have already been fully discussed from the point of view of modifications caused primarily by the light-factor. By such an action of light there is produced, however, not only a change in the individual exposed to it, but a persistent modification, which dies away after the originally modifying factor ceases to act. In this the interaction of the inherited reaction-basis with the external factors is specially clear. The ability to form pigment, and the special arrangement of this pigment—i.e. the establishment of a definite pattern—is naturally a function of the reaction-basis. But the manner in which this function is realized is determined by external factors. In this way arise different colour-varieties which, at the same time, are modifications of the chemical composition as a whole. (Cf. p. 176.)

Their character as persistent modifications is shewn by the following. If, during the critical period just before pupation, orange light is allowed to act on the caterpillars of the Cabbage White, a large number of pupæ are obtained which are poor in pigment (green pupæ). In a particular experiment these formed 62·89 per cent. of the whole number. Following this experiment further, we find that 48·52 per cent. of the descendants of these pupæ if reared in daylight on a grey background shew green modifications—that is, the parental modification appears in a large part of the descendants. That not all of the descendants are equally modified depends simply upon the fact that the environmental conditions we have mentioned favour the production of strongly pigmented pupæ, and that a large number of pupæ yield to this "pressure." Now, in the control material there are usually found, side by side with the normally coloured pupæ, a few green ones (37·2 per cent.). This arises from the fact that a number of caterpillars are unavoidably kept too long, during the critical period, on the green leaves of the foodplant. The proportion of green pupæ in the experimental culture thus exceeds the proportion usually present by 44·8 per cent. There can be no doubt about the existence of a trans-individual action of the original modifying factor, but the nature of persistent modification must be judged by other criteria as well.

It must first be established that the persistent modification observed has not arisen through artificial selection. For it might possibly be thought that only those pupæ which in any case were predisposed by an hereditary factor had been caused to become green by the modifying factor. If this were so, the high percentage of green pupæ would be due to the unintentional selection of this special hereditary factor.

The first fact that argues strongly against this view is that the green and the "not green" pupal colouring are definitely not racial characters, inherited as such, but are always states of pigmentation produced by modification in each separate case.

It is impossible to isolate by selection a "green" race from the whole mixture of races, as could be done in the case of a true race whose peculiarity rested upon inherited factors in the Mendelian sense, and not on modification. The effect on the descendants that we have mentioned cannot, therefore, be one of selection. A second argument, equally strong, is derived from a converse "selection" experiment. If from pupæ reared upon a background favourable to pigment those which are the most pigmented are picked out and their descendants cultured in the same conditions, pupæ are obtained which, in the vast majority of cases, are strongly pigmented (in one particular case, 97·49 per cent.). This might suggest that a race with a tendency to strong pigmentation was being isolated; but this impression is deceptive. For, if it is arranged that the pupation of the progeny takes place in an orange light, there is no trace of this strong tendency to pigmentation. In one particular case 96·01 per cent. of green pupæ were produced—exactly the opposite of what would be expected after selection. No special hereditary factor for the degree of pigmentation is active in this case, and it is therefore impossible by selection to separate such factors into races. If pupation takes place under the influence of a modifying light-factor, such as orange light, the pupal condition of the parents of these individuals is immaterial—that is whether they were pigmented or "green." Whatever the conditions may be, the light-factor modifies the formation of pigment: as we have already said, the state of pigmentation actually attained is always due to modification. If the soma of the parents has been modified experimentally, the progeny will shew, however, in neutral conditions, a change in the same direction—and we shall have a persistent modification.

It is also impossible that here we have to do with a pre-induced modification. If such were the case the original modifying factor would penetrate directly to the reaction-basis, and even in the germ-cell stage would force the course of reaction into a specific channel. Two considerations clearly prove that this is not so. If light, passing through the soma of the generation exposed to it, penetrates to the reaction-basis of the progeny, one should be able by suitable irradiation to modify the progeny even of pigmented pupæ—to make them green, for example. In a particular experiment, pupæ shewing a strong development of pigment, were brought into orange light immediately after they had acquired their pigment. Here they passed the winter; in other words, they were irradiated for many months with orange light. Among the descendants of these pupæ, which were then reared on a grey background, there were found only 2·5 per cent. of pupæ which were poor in pigment—that is, less than is usually obtained in such a culture (see above). The experimental factor therefore remained completely inactive: the attempt to bring about a pre-induced modification had failed.

Finally, it must be stated, in connexion with what has been said, that the primary modification arises only when the eye of the caterpillar is involved in the chain of reactions. In this change we have, in fact, not the direct action of light, but a transformation of the light-stimulus received by the eye; though the nature of the transformation is not yet sufficiently understood. The modifying factor can act on the reaction-basis only after this transformation has been made: that is, after the soma has come into the chain of reactions. Hence the persistent modification obtained is a somatic induction; and, to be precise, it is a hologenic somatic induction. That it is not a merogenic induction follows clearly from the fact that the modifying factor does not directly change the pupal integument, which then passes on the inductive action to the reaction-basis; but, on the contrary, the change in the pupal integument is primarily a consequence of the altered general character of the soma.

To sum up, the origin of this persistent modification can be explained as follows (see Fig. 107, A–E). In the critical period just before pupation the light-factor influences the eye of the caterpillar, and in a way which is not yet perfectly understood this action is transformed and so a special condition of the soma of the caterpillar is produced (107, A). This particular state, which indeed is manifested by a particular chemical composition of the hæmolymph, is communicated to the whole soma of the caterpillar (107, *B*), and this concludes the first phase of the process of modification and induction. During the formation of the cuticle of the pupa in the pupal ecdysis (107, *C*), the general condition of the soma influences the integument of the pupa, so producing a definite kind of pigmentation (shewn in the figure by dots outside the animal). At the same time there

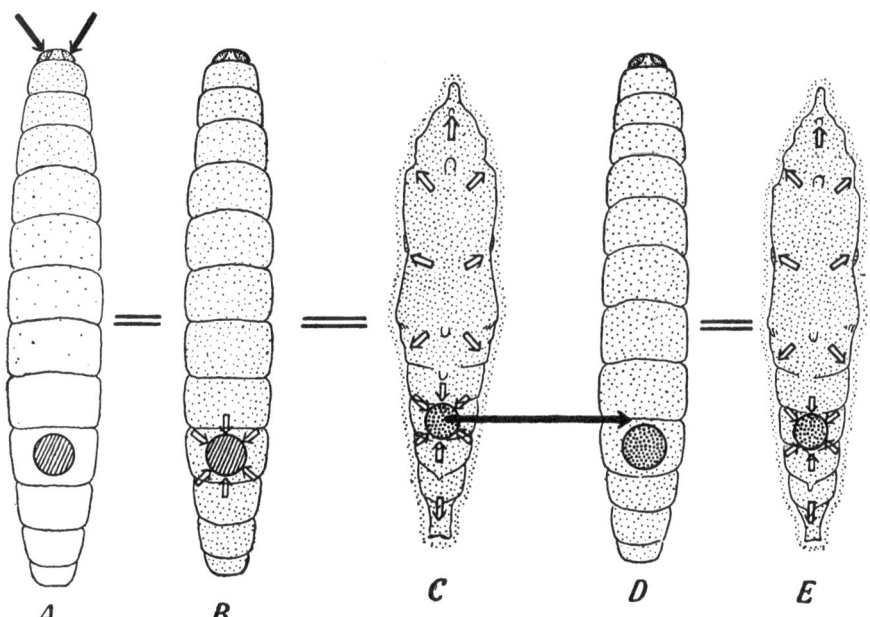

FIG. 107, *A–E*.—Diagrams to explain persistent modification in the pupa of *Pieris* arising under the influence of coloured light. Hologenis somatic induction. Explanation in text.

now begins the second phase of the process of induction, in which the reaction-basis of the next generation, lying in the soma, is affected by the general condition (white arrows). It will therefore be understood that in the third phase of the induction-process this same general condition of the soma reappears (*via* the intervening larval stages) in the pupa of the next generation, modifies its integument in the same way, and again acts upon the reaction-basis, though not so strongly perhaps as in the second phase (107, *E–D*). We see here, then, a form of induction which we have already considered theoretically (Fig. 105, *G*).

(c) *Mutation*

Mutation, that is, a permanent change in the reaction-basis, is by no means a rare phenomenon, though the experimental control of the process is still in its early stages. As an example of directly induced mutation, the increased resistance of *Paramecium* to poisons may be cited. Increased resistance to poisons (e.g. arsenious acid) can arise not only as a persistent modification—in which form we have already seen it—but as a mutation, that is to say, as a really permanent reversal of behaviour towards external factors: not merely an alteration which lasts for a certain number of generations ending with the onset of sexual reproduction, but one which is permanently established in the reaction-basis. In the experiments previously mentioned, such mutations were obtained by the action of arsenious acid, or by raising the temperature during the period of conjugation—though this last result was obtained only during the final phase of conjugation. Clearly the external stimuli must exert their influence during a definite sensitive period, which occurs immediately after the separation of the two conjugants. What may be the nature of the causal connexion between the external factors and the appearance of the mutation, and what the separate processes involved, has not yet been explained. The fact remains that, under the influence of definite external factors, mutations do arise—a result the great importance of which is not diminished by the fact that the mutations, in this special case, are comparatively rare. They are obviously, like the corresponding persistent modifications, directly induced.

This parallelism between persistent modification and mutation is of great interest; it can, moreover, be observed in other cases. The nature of the change here concerned is also notable; for it consists in an increase in the power of resistance to harmful external conditions, so that we may call it an *adaptive* reaction. The specificity of the action should be noted.

Mutations by no means always exhibit this character. They generally take the form of a definite loss to the organism, so that the designation loss-mutation is here a very apt one. A loss-mutation of this kind, whose origin must be ascribed to direct induction, is illustrated by the following experiment in which the trans-individual action of X-rays is responsible for inherited injury to the eye. Mice were subjected to the action of X-rays on five successive days. In the third filial generation, and those which followed, animals were found with abnormal eyes. These abnormalities proved to be heritable, since two pairs of the defective animals produced descendants shewing the same malformations of the eye. At the same time one or more of the extremities might shew club-foot. We are bound to recognize the presence here of a change in the reaction-basis as a result of direct induction; and to conclude that processes of inheritance have actually become involved, without the presence of a Lamarckian inheritance of acquired characters. Since defects of the subcutaneous vascular system of head and limbs were found in embryos, it is possible that the malformations were connected with this primary lesion. To explain this we must assume that the X-rays had injured, in the germ-cells of the irradiated animals, those factor-carriers which take part in the normal formation of the vascular system. If it is assumed that the normal condition of the hereditary factor concerned behaves as a dominant, while in its injured condition it is recessive, it is comprehensible that the injury should not become phenotypic until after several generations—i.e. when two recessive injury-factors come together in the course of inbreeding. The essence of the whole matter is the appearance of a loss-mutation by direct injury to the reproductive cells. Since X-rays are much employed in the practice of medicine, these experiments call for serious consideration from the point of view of a possible danger to succeeding generations.

It is direct induction which is chiefly involved when the germ-cells are damaged by introducing a chemical agent into the soma. As a consequence the reaction-basis suffers a lasting injury, which is manifested as a loss-mutation. As an example of this we may mention experiments on the trans-individual effect of naphthalene poisoning.

For this guinea-pigs were used from a culture the normality and health of which had been established by observations extending over years. A female, mated with its brother, produced, after treatment with naphthalene, two young, a male and a female. The eyes of the male were adversely affected in their development, owing to the action of the naphthalene on the embryo, so that their growth was stunted. This male when mated with a healthy female of the same stock, whose ancestors for years had shewn no abnormality of the eyes, produced nothing but normal descendants. It happened that this F_2 generation consisted exclusively of females. The back-cross of these females with their father, the abnormal male of the F_1 generation, produced six males and three females. One of these latter shewed, on the left eye, the same deformity as the father. Continued inbreeding produced, up to the sixth generation, only females with defective eyes; in the seventh generation there was a defective male which, owing to its premature death, could not be used for breeding. In the tenth generation another male with defective eyes appeared, and by crossing the abnormal females with this male a new race has at last been obtained, which consists entirely of animals with defective eyes. This abnormality appears to be hereditary.

Our commentary on the foregoing experiments is as follows. The injurious substance passed in the first place from the maternal circulation into the embryos, and directly influenced their development; and the defective formation of the eyes is a symptom of this influence. It is clear, however, that some of the primordial germ-cells of the embryos were damaged, so that individual parts of their reaction-basis were specially and permanently affected. In this way there was induced an apparently recessive disease-factor, which, however, was not specifically related to the agent causing the injury. By inbreeding it was possible to obtain this factor as a homozygote, so that the eye-abnormality henceforward appears as a "race" character. In this way a heritable characteristic has certainly been acquired;

not, however, because the original injury to the eye has itself become hereditary—i.e. by the action of merogenic somatic induction—but by the direct liberation of a loss- or injury-mutation.

As we have said before, mutations are not rare. They are most often observed when animals or plants are brought under domestication in any sense of the term. The overwhelming majority of those which so arise can be classed as loss-mutations. A large number of heritable anomalies are already known in Man, and these in the same way are to be regarded as loss-mutations. The different races of mankind and racial characters in general must be taken to be due to mutation. Though all mutations of this, and other, kinds generally arise "spontaneously"—i.e. from causes which are not patent—yet they must have, or must have had, a cause. In the last analysis, this cause must be sought in the action of external factors, even though these factors do not specifically determine the nature of the outward appearance of the mutation. Now, it is more than doubtful whether all these permanent alterations of the reaction-basis are produced by direct induction. For most of the normal factors of the environment are unable to penetrate directly to the germ-cells. We are therefore bound to assume that the soma, whose whole character has been altered by changed external conditions, has here been interpolated; that is to say, we must assume the participation of hologenic somatic induction, even though this be looked upon as merely a non-specific release-mechanism for the most diverse mutations. Seen in this way, the environment—particularly the environment under domestication and, in the case of man, the artificial mode of life inseparable from civilization—assumes a much greater importance. It is essential that the practical geneticist and the so-called eugenist should take this very fully into account.

(d) The "inheritance of acquired characters"

There still remains to be considered the question, already raised, as to how far Lamarckian conceptions are confirmed by the results of experimental research upon the trans-individual action of external factors. Among the typical examples which have been given above there are certainly some whose final result seems at first sight to point to the inheritance of acquirements. But this is only a superficial view.

Now, in the first place, it is often only persistent modifications which are produced—unstable alterations, that is to say, of the reaction-basis; and in the second place we must note that if mutations have, in fact, been liberated by induction, there is present at the same time something other than the inheritance of the induced external character. We must bear in mind that characters, as such, are not inherited at all, and that inheritance consists in the presence of the same reaction-basis in the parents and the progeny. One cannot, therefore, speak of inheritance—to say nothing of the inheritance of an acquired character—if the parents have acquired a new character as a modification, and the descendants have acquired the same character as a mutation; for the new phenotypic character is not inherited at all, and the reaction-basis of the descendants is not the same as that of the parents. Certainly in such a case the same new character has been acquired in both generations, but there has not been inheritance of the genotypic factor for this character—there has been a hiatus in the process of inheritance It is not inheritance, but new-formation. Inheritance is in question only when the later generations shew the same character by reason of the new reaction-basis that they have received. As we have said before, the old problem of the inheritance of acquired characters has now become the problem of the acquirement of a new reaction-basis. This last can certainly be induced by external factors.

On the other hand, there is a large number of phenomena which, from the phylogenetic point of view, it is difficult to regard as other than acquirements which have become hereditary. Many seemingly reasonable proofs of a circumstantial nature can be advanced. The solution of the problems involved must not however be sought along the old Lamarckian lines, but the conclusions which have been briefly stated above must be taken into consideration. To what has just been said on this head, we may add that, so far as our knowledge goes, a true merogenic induction is never the source of a change in the reaction-basis or, at least, has never been proved to be so. This conception of merogenic induction is, however, as we have insisted above, at the root of the principle of historical Lamarckianism. That principle must therefore be essentially transformed to accord with the experimentally established fact that hologenic somatic induction and direct induction are the two sole means by which the reaction-basis can be changed. A particular change of a part of the soma cannot therefore—as the original Lamarckianism would make it—become directly heritable. At most it can become heritable indirectly by way of hologenic somatic induction, in which a change in one part of the soma first alters the whole, after which there is the possibility of a corresponding change of the reaction-basis by induction.

This means, however, a considerable limitation of the sphere of application of the remodelled principle: for not every change of the soma, or part of the soma, which is produced by external factors or by function and the like, comes into question as the starting-point of a phylogenetic transformation. Only a change involving the whole character of the soma can act in this way.

We cannot further discuss the problem of evolution here. This one conclusion may be drawn: What is proved by the exact investigation of the trans-individual action of external factors is not historical Lamarckianism and its traditional conception of the inheritance of acquired characters. What is proved is simply that the question remains open as to whether by hologenic somatic induction, as here defined, there can arise trans-individual advances in development which are also phylogenetic. The answer is probably an affirmative one.

CHAPTER VII

THE VARIOUS INTERRELATIONS CONCERNED IN DEVELOPMENT, THEIR FUNCTIONS, AND THE AGENCIES UPON WHICH THEY DEPEND

I. Interrelations Connected with Determination

1. The Subordination of Separate Interrelations within the Whole

The manifold relations which exist between the parts of the germ, and between the germ and its environment, may be divided into two main groups: those which *determine* and those which *realize;* and in each of these groups different categories can be distinguished.

In these relations of the parts to one another, and of the whole germ and its parts to the environment, there are various points of interest to be considered: in particular, the actual thing which is done, the agent active in its performance, and the way in which this agent acts. The *formal* aspect of the relations—whether, that is to say, we have to do with the simple dependence of one thing upon another or with a reciprocal dependence—this also claims our attention.

As to interrelations which are determinative, these are appreciable only in so far as they involve the interdependence of visible parts of the germ; and this is so whether they be present in the egg before cleavage or only in the germ after its division into cells. We can come to no conclusion at present about the constitutive, non-morphological aspect of the reaction-basis by investigating such interrelations.

In considering the determinative interrelations within the germ, it must not be forgotten that determination as a whole is a function of the whole germ, and not simply the sum of the actions of its morphological subdivisions. Attention has already been drawn to the fact that the fate of a part in development is always decided "within the framework of the whole" and with reference to the whole. Not only is there separate activity of the parts, but there is also an action of the whole, since development is epigenetic—i.e. is not merely the unfolding of a diversity that is already present in its entirety, but the production of a manifoldness that is really new. To say that determination as a whole is the sum of separate interactions would be to abandon this result of the analysis of development, and return to a purely preformistic conception of development. In view of what has been advanced in the earlier part of this book, that is not possible.

The origin by epigenesis of the various degrees of advance in organization is essential to development. The interrelations of the different germinal regions which arise in this way are partly determinative in character.

Even before the formation of the subordinate organizations to which we refer there are, of course, different parts present in every germ-cell. The cell shews differentiation, for example, into nucleus and cytoplasm, and more detailed differentiations which are—morphologically—exhibited most clearly in the nucleus. Among these primary parts there are certainly interrelations which assist in the first steps of determination. We cannot be more precise in detail, but can only affirm with certainty that the whole of determination does not proceed from a single primary part. Again, with reference to the primary parts and their importance in determination, it must be reaffirmed that, in some way or other, these parts are subordinate to the entirety of the germ, for this is demanded by the epigenetic character of development. The ultimate foundation on which determination as a whole rests is the entire specific constitution of the reaction-basis: the separate parts—both those initially present and those which appear secondarily by epigenesis—work within the limits imposed by its activity. What is present at the beginning of development does not consist simply of rudiments in which everything that is to be produced is already preformed; and determination taken as a whole does not consist solely in the activation of such discrete rudiments. It would be more true to say that determination was the activation of potencies—the liberation of developmental possibilities.

There arises here a problem which cannot be regarded as solved—the problem as to whether determination is, so to speak, a negative or a positive process; whether the more or less exclusive laying-down of the fate of a particular region consists in the inhibition of all except some few potencies, or whether in the activation of these few potencies, the majority remaining latent. Obviously, also, both kinds of processes may occur at the same time. It has been pointed out that the process of determination does not, in all circumstances, create at the outset a rigid, unchangeable situation—even leaving out of account the phase of *institution*, which is always subject to change. The existence of heteromorphic regeneration shews that other rudiments and potencies than those responsible in ontogeny for a particular region can become active after the close of primary development. This might mean that determination during ontogeny consisted in the temporary inhibition of potencies. When, again, the individual blastomeres of the egg of *Tubifex* have their fate decided by the removal of pole-plasm, it looks as if an inhibition of potencies took place. On the other hand, the organizing action of the blastopore lip is certainly to be conceived as a positive one. Still, it cannot at present be decided whether the result of this action is brought about by certain parts of the reaction-basis being set in motion before the others, or simply by a putting out of action of the rest of the internal factors of development. The solution of this problem, which must be left to the future, is of the utmost importance for our definition of the essential nature of determination.

2. Agencies Concerned in the Determinative Action of the Parts

As to the agency by which determination is carried out, this cannot be considered to be exclusively material in nature when we look at determination as a whole—though, of course, material agents are often concerned, since it is a material germ with whose development we are dealing. In part, however, determination must depend on the action of agents which are not material or chemical. However surprising this may seem at first, there are weighty reasons for the opinion. In the first place there is the well-established conclusion that development as a whole is epigenetic. Now, if the agents of determination were purely material, their action would have to be regarded as purely chemical or as involving also chemical stimulation; and it would be impossible to avoid the assumption of a complete preformation of all the parts, since such actions could only be the result of agents which were pre-arranged from the beginning. Again, there are things which occur in development which cannot be explained on the assumption of purely material means of determination, though these things have as yet hardly come within the scope of the analytical study of development. What especially defies this kind of solution is the problem of the typical *facies* and form of the whole organism and of its parts. For there must be produced in development not only definite chemical substances, "ear-marked" for each of the tissues of the adult organism, but also a specific facies or configuration of the parts and of the whole formed from such substances. And the configuration itself must be determined.

It seems impossible that this part of determination should be effected by purely material means. Hence it is necessary to admit the possible agency of what may be called energy-relations, about the character and action of which we can at present, however, say nothing with certainty. There is, at the same time, no lack of experiments on which definite ideas may be based: the conception, for example, of the morphogenetic or embryonic *field*—the idea of a field being borrowed from physics (cf. electromagnetic field). What is imagined is, to put it simply, a space containing a definite arrangement of energy-conditions: a totality made up of all the diverse forces, varying in kind and in magnitude and in direction, which act upon the separate points of the field. In this aspect, then, the living system consists of the visible germ, and at the same time of a field which cannot be perceived by the senses but whose existence can be indirectly inferred. It is in this field that the events of development and of morphogenesis take place. Hypotheses which place in the foreground what is purely "energetic" and dynamic are perhaps destined to influence very greatly all our notions of what happens in development. For the present they are chiefly of value as indicating an increasing recognition of the impossibility of assuming a purely material agency in determination.

From the study of individual processes of development and the relations of these to particular constituent parts of the germ it appears—more clearly than from such general discussions— that these separate processes participate, sometimes to a marked degree, in the determination and realization of developmental fate, but that this participation is without prejudice to the importance of the whole.

We have already given examples of the determinative relations of particular germinal regions to one another, as in the case of inductive and organizing action in the various stages of organization. We must now consider briefly what are the agencies by which those relations are brought about.

In dealing with these activities of the parts of the germ it must be borne in mind that there are two kinds of agencies— those which are material, and those involving energy-relations only. In the case of material agents it must further be settled whether these act directly as chemical agents, or whether they are the carriers of chemical stimuli, without entering into real chemical reactions with the regions which they influence. In the latter case they would occupy an intermediate position between the purely material and the purely "energetic" agents.

As is well known, the point of view taken in genetics—and on good grounds—is that sex is an inherited character, and that what decides whether a male or a female shall appear is the occurrence of a disjunction like that in a Mendelian backcross. The heterochromosomes (heterosomes)—appearing as the X- or Y-chromosomes—are recognized as the carriers of that which causes sex. Two types can be distinguished among animals for which this generalization holds good. In the one type to which belong Mammalia, Orthoptera, and Diptera, the female possesses, as the result of a particular combination when the egg is fertilized, two X-chromosomes; the male, on the other hand, has an X- and a Y-chromosome, though sometimes the Y-chromosome is lacking. In the other type, to which belong Birds and Lepidoptera, the converse relation subsists: the female contains only one X-chromosome, and the male two. The first type we may call the Drosophila type (from *Drosophila melanogaster*, a fruit-fly), and the second the Abraxas type (from *Abraxas grossulariata*, the Gooseberry Moth), because these were the animals chiefly used in the experiments which demonstrated this behaviour. Inasmuch as the material constitution of the nucleus, depending upon the chromosomes present, affects the determination of sex, it is obviously possible that this determining action itself is brought about by material means. Either the sex-determinants in all the cells of the body might produce special substances responsible for the sexual bias of these cells, and therefore of the whole body; or, as a possible alternative, material messengers of the determinants might desert the cells and distribute themselves throughout the whole body. What cannot be decided, however, is whether these material agents of determination act directly as chemical agents by entering into their own proper chemical reactions with the other substances of the cell, or whether they are simply the carriers of chemical stimuli to which the living cells respond by a definite adjustment of their developmental fate.

In early Amphibian embryos, material from the young medullary plate can induce a medullary plate in its immediate neighbourhood, a phenomenon which has been called *assimilative induction*. This relation, again, suggests the presence of a material agent; that is, that specific substances arise from the inducing material, and that these—either as carriers of stimuli or as chemical agents—influence developmental fate in their immediate neighbourhood.

An attempt has indeed been made to explain the action of

organizers by the assumption of a material connexion of this sort, even in cases where the product of induction is different in character from the inducer. It is at the moment impossible to decide this question; for there might well exist energy-relations capable of effecting the determination of one germinal region by another acting as organizer. The object of this discussion is not, however, to give a complete explanation of determinative interrelations by shewing what agencies are active in them, but only to call attention to the number of important problems that arise in this connexion.

Now, there are cases in which something—though it be only of a negative sort—can be predicated of the agents active in determinative relations; and in the particular cases in point the observed interactions cannot be attributed to material agency. From among the many examples of determinative relations already mentioned, and which might be quoted again in this connexion, we may call special attention to two.

It has been explained in detail how it was possible to obtain a typical limb from the regeneration-blastema of a newt's tail, if the blastema were transplanted sufficiently early in the neighbourhood of the normal fore-limb of the animal. Unquestionably the fate of the transplant is here influenced by what lies below it. Now, it is inconceivable that this is brought about by a material agent. If that were the case, formative substances would penetrate from the substratum into the blastema, and it is not easy to see how an ordered whole could in that way be produced. For these substances would not encounter an arrangement of the material system appropriate to a limb—as we must suppose them always to do in a normal limb-bud—but any arrangement that they did encounter must certainly be one appropriate to a tail. In the absence of such an arrangement, the production of a whole, typical in its form and in the ordering of its parts, is inconceivable. It could certainly never be produced by substances diffusing in. We are forced therefore to imagine an energy-relation as the agency in determination; and of this, perhaps, we may form some idea if we look upon it as the emanation of a morphogenetic field.

Of the nature, direction, etc., of the energies or forces that might be concerned nothing can at present be said.

From the present point of view the following experiment on the earthworm (*Lumbricus rubellus* and *Allolobophora terrestris*) is instructive. From the hinder region of a worm a piece is taken, the anterior end of which would, normally, regenerate a heteromorphic tail. This piece is grafted side to side on to the anterior end of another worm, so that transplant and host grow together lengthwise (Fig. 108). While the piece used as a transplant always regenerates, when free, a heteromorphic tail on its anterior cut-surface, the union which we have described with the anterior end of another worm inhibits this regeneration.

Fig. 108.—Earthworm with transplant parallel to the anterior end of its body, the transplant having been taken from the hinder region of another worm. (After F. Mutscheller.)

Corresponding inhibitions are also seen when, after such transplants are healed, a cut is made to divide both host and graft transversely. In that case the regeneration of the head-end of the host is also inhibited. Besides the quantitative influence of the substratum upon the regeneration of the graft (and the reciprocal effect), there is also a qualitative influence on the transplant. Thus, the better the union of the transplant with the substratum, the more head-like are the structures which arise on the cut-surface of the former. It is not a matter of indifference whether the transplantation is made side-to-side or back-to-back; for it has been shewn that the nature of the union has an influence on the form taken by the regenerate arising on the transplant. It is unnecessary to go into further details.

Though, indeed, determining influences are not the only ones which are here at work, such influences are certainly active side-by-side with those which produce realization. Any attempt to refer them to material agency fails, as is proved by the following considerations.

Let us assume that in the anterior part of a worm a specific substance A brings about the regeneration of a head-end, and in the posterior part a substance B, the formation of a regenerated tail. Then it seems at first possible to explain inhibition of the regeneration of the transplant in the above experiment as follows. The substance A passes from the substratum into the transplant, and comes into conflict there with the substance B. This might result in the suppression of the tail-regenerate; or, in the experiment in which there was a cut-surface common to both transplant and host, the two substances might be mutually injurious. Against this, however, must be set the fact that the form of the structures which actually arise depends on the degree to which transplant and host have grown together; and this cannot be explained by such an interchange of substances. Apart from other considerations, however, material influences are quite inadequate to explain the fact that the nature of the union of the two components— whether side-by-side or back-to-back—affects the quality of what is formed; for the nature of the union can neither help nor hinder the exchange of the hypothetical substances, since in both cases the union is extensive and intimate. Here, then, it is only by assuming a non-material (i.e. "energetic") agency that we can explain the relations in question. One is tempted again to think of the morphogenetic field which has already been referred to; but it is sufficient for our present purpose to have demonstrated in this case also the impossibility of a causal connexion which is material in character.

Further examples of such non-materially conditioned inter-relations can be found by the reader himself among the experiments described earlier. Anything that can be said on this subject must still be very speculative in character, as also, indeed, must be anything said about the relations which are materially conditioned; but the problem has been opened up, and will probably play an important part in the investigations of the next few decades.

Something may be added in this connexion. It has been clearly demonstrated that growing and embryonic tissues give rise to radiant energy. This energy, given out, for example, by the embryonic brain of the Amphibia, the early stages of the chick embryo, and by growing onion-roots, has the peculiarity of being able to induce nuclear division in other tissues. We speak, therefore, of *mitogenetic* rays. The discovery of these phenomena has alone contributed much to the establishment of the conception of the embryonic field. The influence in question does not possess a true determinative character, but at the same time it is very interesting to find that good evidence can be advanced to prove that energy is indeed radiated from parts of the germ.

Since in determinative relations a part is played by stimuli—whether these be purely "energetic," or whether chemical and due to the instrumentality of specific substances—one may speak of structure-stimuli. In development these stimuli vary in importance and in kind. Those which here interest us most stand in immediate relation to the quality of what is formed, since they enter into the process of determination. On account of the part they play in morphogenesis they may be called *formative stimuli*; but we must remember that there are other stimuli which though they are important in morphogenesis, are not themselves determinative or formative, and therefore do not directly belong to determination. In those stimuli which are active in determinative relations the specificity of the stimulus is all-important, while in the others which remain to be described this specificity is not of decisive importance.

II. INTERRELATIONS CONNECTED WITH REALIZATION

When the determination of a region is completed its later development has by no means been settled with certainty. Before its fate is further qualitatively established there must still take place many processes of realization. In these, again, numerous important interactions come into play, so that in contradistinction to interrelations and interdependences which are determinative we can speak of those whose function is to *realize*. There is not, of course, in actual fact, this sharp line of division between the relations of the parts in development —it would be nearer the truth to say that the same state of interdependence may manifest itself in several aspects according to circumstances, and that a diversity of interrelations and interdependences are connected and cross-connected in the most complicated fashion. But for the purposes of analysis such complications may be left out of account, and the relations involved in realization may be considered separately as a special kind of interdependence of the parts, for in many cases such a discreteness is indeed observed to exist.

In these relations of realization we encounter once more the actions of stimuli, and since these stimuli are chemical and physical it is with mechanistic agents that we have to do. As regards the stimuli involved, we may first distinguish those whose action is *augmentative* and *trophic*. In this case what is directly brought about by the action of the stimulus is only a quantitative change, in which the stimulus can have both an initiating and a maintaining or promoting (protective) action— we are in fact dealing with a growth-stimulus. In the second place there can be distinguished stimuli whose action is *formative* and trophic. Here the stimulus indirectly assists in the production of structures which differ qualitatively. It does so either by a specific local action which makes it responsible for the character of a definite region, or, in the form of an executive stimulus, by actually making possible the process of development. In such cases it is generally executive or control-stimuli that are present.

Among the interactions connected with realization we must number the various ways in which development is dependent upon environmental factors. Any such kind of dependence may fall into one or other of the groups of stimuli mentioned, but from other points of view may shew peculiarities. For, on the one hand, they may be a source of variation; on the other hand, we may find in them not only the action of stimuli but that also of direct chemical and mechanistic agencies. Chemical intermediaries condition the dependence of one part upon another in internal relations of realization. Specific substances certainly act as carriers of stimuli—as, for example, in the case of the hormones of the endocrine glands, of which more will be said later. For the rest, it is not always easy in dealing with these interesting phenomena to separate the determining from the realizing influences, and these two sides of the relation may often be in fact inseparable.

1. The Augmentative-Trophic Relation

The purely augmentative-trophic aspect of an interaction or of the stimulus concerned in it consists simply in the increase in size of a part or tissue already present. In normal embryonic development such actions are certainly very important, but they cannot, for the reasons already stated, be distinguished separately. They appear, however, very clearly in the advanced stages of formation of organs and tissues. The quantitative effect of the functional stimulus on muscle-formation is an example. A similar phenomenon occurs in the functional hypertrophy of the kidney or the liver, which appears when one kidney or a part of the liver is extirpated. The single kidney, or the remains of the liver as the case may be, must now perform the same function as the whole organ. This increased use provides the stimulus which brings about the hypertrophy. In the case of the kidney and the liver what are active are chemical stimuli dependent on the character of the blood entering an organ that removes certain of its constituents. Again, partial extirpation of the thyroid gland leads to a functional hyperplasia of the remainder. No qualitative changes of any kind are produced by such processes. Whether one regards histological character or general structure, only a simple quantitative action is seen.

2. The Formative-Trophic Relation

(a) *Interrelations conditioned by stimuli*

Formative-trophic relations can be demonstrated in ontogeny and in regeneration much more clearly than can augmentative-trophic relations; and their several functions are more various. In such cases, as we have pointed out, the existence of this relation is indirectly of importance for the quality of what is formed. A stimulus active in this way has not itself a qualitative, specific action, since the quality of the product depends on entirely different things; the quality of the result is, however, involved in the nature of the interference of the stimulus. The following examples will illustrate this in detail.

By the addition of various potassium salts to sea-water the formation of the characteristic calcareous skeleton of the sea-urchin larva (Pluteus) can be prevented. The normal pluteus possesses, as is well known, long, arm-like projections in which lie the thin, calcareous skeletal rods (Fig. 109). When the formation of the rods ceases, the development of the arms is suppressed. Clearly the calcareous spicules exert on the tissues a stimulus which causes the development of the projections, and if the stimulus is lacking the formation of the projections does not take place. This in itself is not a qualitative stimulation: obviously a simple growth-stimulus is alone concerned. If, however, we take into account not only the direct action of the stimulus but the structure of the larva as a whole, it must be admitted that the action of such a local stimulus creates something that is indeed new in quality. There is formed in fact a larva with projecting arms, while in the absence of the stimulus a larva without arms arises, distinguishable from the normal larva not only by its lesser size but by a completely different *facies*—i.e. qualitatively distinguishable.

Though here it is the quality of the external form which is conditioned by localized stimuli—which must be regarded as growth-stimuli—there are also other cases in which a more special structural condition arises in this way. The way in which the structure of spongy bone depends on the mechanical demands made upon it furnishes a positive proof of this, as we have already shewn. As far as the tissues present are concerned, that which arises under the influence of the stimulus is not new; but the arrangement of the tissues in the architecture of the bone results in a distinction between the bone in question and

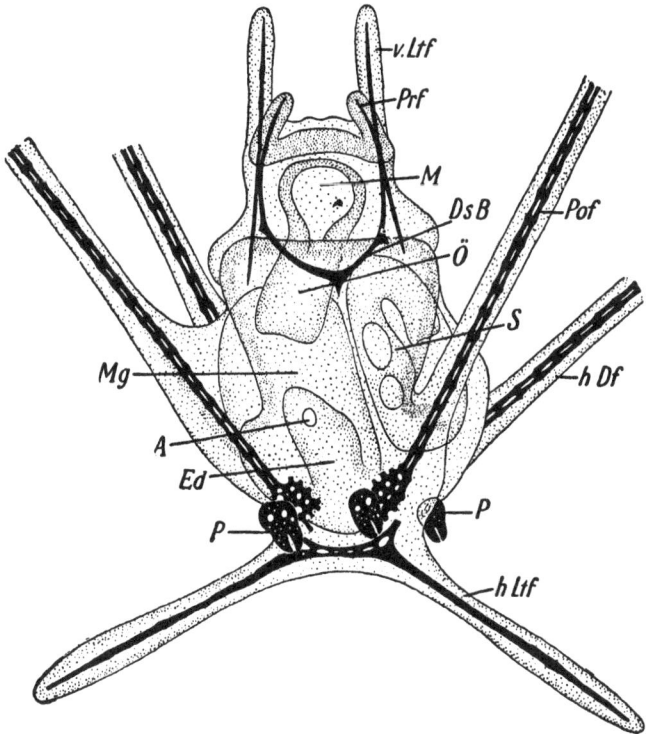

FIG. 109.—Pluteus of *Arbacia pustulosa*, ventral view. *M*, mouth; *O*, œsophagus; *Mg*, stomach; *Ed*, intestine; *A*, anus. Processes ("arms") and skeletal rods: *vLtf*, antero-lateral; *Prf*, pre-oral; *DsB*, dorsal loop; *Pof*, post-oral; *hDf*, postero-dorsal. *P*, pedicellaria; *S*, embryonic rudiment of the body of the sea-urchin. (After von Ubisch.)

homologous bones loaded in a different direction. This distinction is not quantitative, but concerns the quality of architecture.

To the group of formative-trophic relations belongs the interdependence of the peripheral organs and the nervous system. This can be demonstrated in the development of the frog.

If in the very early larva of *Rana fusca* (*R. temporaria*) the rudiment of one hind leg is extirpated, considerable developmental inhibition is produced in various places. First, half of the pelvic girdle of the side in question does not develop; again, the transverse process of the sacral vertebra, with which, normally, the wing of the ilium articulates, remains weak and short, so that it is indistinguishable from the transverse process of any other vertebra. Further, the nerves of the leg and their centres in the spinal cord are backward in development. These inhibitions can extend so far forward along the central nervous system that even the brain may shew developmental lesions. In the peripheral and spinal nervous systems the inhibitions produce not only a lesser development of the nerves proper to the absent leg, of the spinal ganglia, and of one half of the spinal cord, but there is a reduction in the number of cells in these ganglia and in the anterior horn of that part of the spinal cord which is concerned. Where the inhibitions pass as far forward as the brain, it is especially the median part of the roof of the mid-brain which is involved. Moreover there is in this place not merely a reduction in size but an actual arrest of differentiation. This is best seen by comparing transverse sections through a normal mid-brain and through one that has been inhibited in this way. Fig. 110 shews, on the left, a section through a normal mid-brain; on the right, another section through one that has been inhibited; comparison of the two shews their differences to be very obvious. The thinning of the roof is due not solely to the weaker development of its various layers, but to a suppression of the complicated stratification of the normal brain, as is plainly apparent in the figure. Extirpation of the rudiment of a fore-limb gives comparable results.

Inhibitions of development also occur in the mid-brain if the eye is removed from a young larva. The brain-centres for the missing eye shew a considerable reduction, so that a cross-section through such a brain gives an asymmetrical picture (Fig. 111).

In connexion with this kind of developmental inhibition of

the nervous system after removal of a limb-bud or of an eye, abnormalities have often been observed in the limbs which were not operated on. These abnormalities, too, consist in inhibitions not of growth alone, but of the differentiation of the limbs. These inhibitions may vary in degree. Either the distal portions of the extremities are alone affected, so that the digits are defective; or whole segments of the limbs are affected by the inhibition, and there is no approximation to the normal

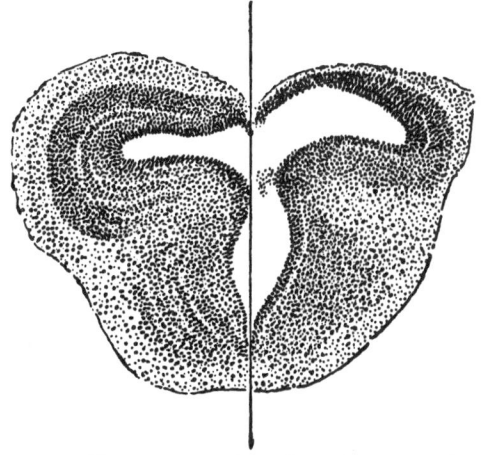

FIG. 110.—Transverse sections through a normal mid-brain of *Rana fusca* (*R. temporaria*) and through one with its differentiation inhibited (as a result of early extirpation of the left hind-limb rudiment and the concomitant inhibition of the limb of the other side). Half of each section is shewn for comparison: left, normal; right, inhibited.

form. Here again, as in the case of extirpated limbs, the development of the nerves and spinal centres of the inhibited limbs is adversely affected. An inhibition (at least of the distal segments) of the limb, has been experimentally produced by removal of one optic lobe—that is, one half of the roof of the mid-brain—from the young larva.

A completely satisfactory and detailed explanation cannot yet be given of the mode of origin of these inhibitions. The following is a possible explanation. As a result of the early loss of a limb the peripheral nerves, spinal ganglia, and regions of the spinal cord associated with the limb remain backward in development; and this inhibition exercises an influence upon the centres higher in order—passing from neuron to neuron— so that finally the controlling regions of the mid-brain are inhibited. Where the inhibition is particularly intense it reaches and affects, in the mid-brain, the centres for the unoperated limbs and the opposite side, these centres being connected with those first mentioned by the function of correlation. There now takes place an inhibition passing down from the mid-brain to the limb-regions of the spinal cord, the spinal ganglia, and the peripheral nerves of the limbs which were not

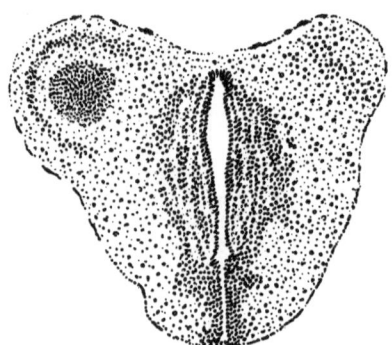

FIG. 111.—Transverse section through the anterior region of the mid-brain of *Rana fusca* (*R. temporaria*). Inhibition of one side resulting from the early extirpation of an eye.

affected. These last, in response to the inhibition of their nerves, now shew an inhibition of their own development. Something quite comparable would be true of an inhibition passing down to the limbs after extirpation of an eye, and this because the mid-brain is affected in a very similar way whether the extirpation be that of a limb or of an eye.[1]

Now, it has recently been demonstrated that the limbs of the frog can develop quite normally even when they are primarily aneural, that is if the connexion of the limb-rudiment

[1] In my *Lehrbuch der Experimentalzoologie*, 2nd ed., 1928, this explanation is still given, since when it was written the existence of limb-development in the absence of spinal nerves in the limb-rudiment had not been confirmed. This explanation must be abandoned, however, since it now appears that, in the frog, limbs that are primarily aneural can develop normally.

PLATE XI

a   *b*

FIG. 112, *a* and *b*.—Mexican Axolotl: *a*, ordinary form with persistent gills; *b*, adult form some months after metamorphosis brought about by feeding on ox-thyroid (both figures from living animals). Note in the metamorphosed example—besides the absence of gills—the differences in shape from the larva, particularly as regards the head and tail, and the stouter build of the limbs. The slight and often irregular web which is generally present between the digits of the larval hand also degenerates.

with its spinal nerves is from the outset experimentally prevented. We must therefore look for another explanation of these complex phenomena.

It is quite clear that the developmental inhibitions which pass up to the brain are causally connected with absence of the peripheral organs, and that these inhibitions arise in the way described. Descending inhibitions, however, cannot pass by way of the spinal cord and peripheral spinal nerves because, as has been said, normal limbs can arise without connexion with the spinal cord and spinal nerves.

Whatever further investigation of the individual connexions may yield, the formative-trophic interrelations of the peripheral organs and the nervous system, as they appear in these experiments, are of great interest, and shew that these relations form an unusually complicated system. The agents responsible for them are certainly, in the main, direct stimulations. We can, indeed, recognize here an *executive* or *control-stimulus*, which, though it has itself nothing to do with the determination of the regions concerned, is partly responsible for the quality of the thing formed, by reason of the way in which it acts in the process of realization.

In the formation of the corneal epithelium (conjunctiva) of the Vertebrate eye, determination and realization are closely associated, as is shewn particularly well by experiments on the Amphibia (cf. above, p. 152). The typical differentiation of this epithelium depends on the eye. It has been attempted to refer this action to a specific substance, produced by the eye as an internal secretion, the substance being supposed to act on the skin only in the neighbourhood of the eye, since only thin layers of connective tissue are permeable to it. Whether this assumption will be confirmed remains to be seen. A substance of this sort should not be compared to hormones, for these are diffused throughout the whole body. Moreover, in order to explain the stimulation of living tissues we are not entitled to ascribe gratuitously a specific secretion to every organ and tissue.

Various aspects of this relation can easily be distinguished. Apart from the purely determinative component which is

P

manifested, we see that the direction taken by the differentiation of the skin is here dependent on the influence of the eye; and there is thus present at the same time something which directs and something which maintains. For as soon as the influence of the eye ceases, the process does not merely come to a standstill, nor is the stage of differentiation that has been reached maintained, but regression sets in. The stimulus concerned has therefore the character of one which controls and maintains.

The importance of an executive stimulus is beautifully shewn in regeneration, the dependence of the completion of which upon the nervous system has been demonstrated in Vertebrates, at least in some cases. For example, if a limb of *Triton* is amputated, and if at the same time the nerves to the stump are cut, regeneration certainly sets in, a conical blastema being formed; but the regenerate remains extremely incomplete. Elimination of the whole of the spinal apparatus concerned—by cutting out the appropriate piece of the vertebral column—suppresses the process of regeneration. If the regeneration is allowed to begin and the nerves are then cut, the process comes to a standstill. If the nerves regenerate, the regeneration of the limb can begin again. The question as to which components of the nervous system are here active has received no unequivocal answer. Much can be said for the view that it is the sympathetic nervous system whose action produces the controlling stimulus. This is supported, first, by the fact that it is not the quality but only the intensity of the regeneration that is influenced by the nerves, and no specific morphogenetic influence is therefore at work, but only a general stimulation which is in keeping with the action of the sympathetic. There is the further evidence that if the sympathetic cord alone is cut out, without injury to the motor and sensory centres of the spinal apparatus, the regeneration of the limb ceases in the same way. In any case the sympathetic appears to be competent to start and to maintain regeneration unaided. Besides the controlling stimulus, an initiating stimulus is thus shewn to be concerned in these phenomena.

Now it is interesting to remark that an initiating stimulus of this sort can be isolated for observation. In the Hydroid, *Eudendrium racemosum*, the formation of polyps is dependent upon light. In the dark they are either not developed at all, or are developed in very small numbers. The formation of stolons, on the other hand, is independent of light. Blue rays have a specially favourable action; red rays have the same effect as darkness. Light is also of importance in the regeneration of *Eudendrium*. Hydranths are regenerated only in light, while the regeneration of stolons is not prevented by darkness. This dependence of the regeneration upon light may indeed be masked by the fact that there has been a temporary exposure to light—which suffices to start the new formation. If the hydroids are kept in darkness for no longer than thirteen days, polyps are just as well regenerated as in light. They are still under the influence of the previous exposure to light—which appears from the fact that after longer sojourn in darkness regeneration does not occur. Exposure to light for only a few minutes, however, is sufficient to start regeneration; but this exposure is absolutely necessary. From that time onwards, regeneration is independent of light. Here the function of the stimulus is that of initiation; it is not a condition of the carrying-on of the process.

(b) Interrelations conditioned by hormones

In the majority of the relations so far described we have dealt with stimuli whose action is direct; but there also exist kinds of interdependence in which the stimuli are transmitted by definite substances. Such carriers of stimuli are found principally in cases where the interrelated parts are widely separated from one another, so that a direct relation between them would be out of the question. If, therefore, a dependent relationship exists, this must be through a special intermediary. The substances known as internal secretions or hormones are to be regarded as carriers or intermediaries of this sort. By hormone is meant, strictly speaking, the product of an endocrine gland. Hormones are principally of importance among Vertebrates; in the invertebrates they clearly play a very subordinate part, though a few exceptional cases are known.

Among the hormone-producing organs of the Vertebrates are the thyroid gland, the parathyroid (epithelial bodies), the thymus, the pituitary (hypophysis), the epiphysis, the adrenals (supra- and interrenals), and the gonads. There exist many interrelations among these various glands, so that together they form a correlation-system, the components of which act upon one another in the most complicated way. This can be shewn by extirpating individual glands and studying the resulting phenomena in those which remain. From this point of view pathological observations are of great importance. There is a specially close connexion between the gonads, the thyroid gland, the thymus, the pituitary, and the adrenals. It is the hormones of the glands concerned which are responsible for these relationships. The picture is further complicated by the fact that many morphogenetic processes throughout the whole body depend upon the secretions of particular glands, so that investigation of the hormonic correlation (and relation) enables us to visualize the arrangement of the individual parts in the whole.

The connexion between developmental processes and the endocrine glands is demonstrated by pathological cases in which there is a disturbance of the source of some particular hormone. In the case of the thyroid gland, the feeding of Amphibian larvæ with this organ shews that its hormone hastens differentiation and metamorphosis. Such feeding increases the amount of the active hormone present. Body-growth decreases, and this is probably connected with the facts that increase in the thyroid hormone limits the intake of water, and that the great loss of water by the tissues before normal metamorphosis may be attributed to increased activity of the thyroid gland. Even species which normally exhibit a greatly retarded metamorphosis, or which like the Axolotl do not generally change from the larval to the adult form, are made to metamorphose by being fed on thyroid gland. As is well-known, the axolotl becomes sexually mature as a larva, and retains its external gills throughout its life. On feeding these animals with thyroid gland they lose their external gills and become adults, leaving the water and living henceforward on land (Fig. 112, *a*, *b*).

Extirpation of the thyroid gland leads to inhibitions of development. Amphibian larvæ without the thyroid gland develop normally up to the time of the formation of the limbs; but then complete inhibition of development occurs. Different organs react unequally to alterations in the amount of the thyroid hormone, so that a disharmony of development results. Malformations and typical mass-displacements can arise by upsetting the hormone-balance in a purely quantitative way.

In placental mammals the thyroid hormone of the mother may, during the fœtal period, be important to the offspring too. Certainly in man cases of inborn thyroid deficiency are known, in which the actual effect of the deficiency appears only after birth; it must therefore be assumed that during the fœtal period the maternal hormone has compensated the deficiency. Lack of thyroid in man produces myxœdema, idiocy, and cachexia. Investigations on sheep have shewn that extirpation of the thyroid gland leads, a little while after birth, to strong inhibitions of growth. Since thyroid deficiency affects the other endocrine glands, changes in these are, in certain circumstances, probably responsible in part for developmental reactions in the body.

Increase in the amount of thymus hormone by feeding tadpoles on this gland leads to a great increase in the size of the body, and at the same time a postponement of metamorphosis, which indeed is finally suppressed. On the whole, then, this influence opposes that of the thyroid secretion. In adult mammals extirpation of the thymus is without result; but in young animals this operation produces pathological deficiencies in development and growth which are similiar to those caused by thyroid deficiency. Statements have been made which do not accord with what is here said with regard to the role of the thymus secretion, but in any case it is an important one.

The relation of the pituitary, and especially of its anterior lobe, to growth-processes is shewn very clearly in the phenomena of acromegaly and gigantism. Acromegaly appears in connexion with a hyperplasia of the pituitary in animals whose growth has ceased, and consists in an abnormal increase in size of the extremities of the body (hands, feet, skull); gigantism, in young individuals, leads to elongation, especially of the limbs, the gonads at the same time being adversely affected.

Extirpation of the pituitary arrests the development of young frog larvæ—which, however, begins again after implantation of the anterior lobe of an adult frog. An increase of the secretion, caused by feeding metamorphosed frogs on anterior lobe, produces giant growth, a result which accords with the pathological observations mentioned. All the details of the action of the pituitary hormone have not yet been elucidated, in spite of numerous experiments; and this is also true of the other endocrine organs.

Pathological observations indicate that the secretion of the adrenal cortex in mammals is related, amongst other things, to sex-differentiation and to secondary sexual characters. Precocious sexual maturity, with premature development of the secondary sexual characters, is generally connected with a hyperplasia of the adrenals, and in particular with an adrenal tumour. Since the gonad generally appears to be normal in such cases, there is obviously a causal connexion between the change in the adrenals and the abnormally early appearance of the secondary characters; but possibly the connexion is more indirect and involves the gonad hormone. Experimental investigations into the significance of this secretion have not yet yielded consistent results.

The action of the gonad secretion is that which has been studied most thoroughly, partly because it exercises so far-reaching an influence on the whole body. In the invertebrates this is by no means so. Cases are indeed known in which the gonad hormone plays a definite part, as in the Crustacea; but there are large groups of invertebrates in which nothing of the kind is found. This last holds good for the Insecta, where the secondary sexual characters have proved to be completely independent of the gonads; but in the Crustacea such a dependence exists, at least to a limited extent.

In all the groups of vertebrates, on the other hand, the gonad hormone is of capital importance. This hormone influences not only the actual secondary sexual characters, but the whole make-up of the individual with regard to characters unconnected with sex, such as the general condition of its skeleton and musculature, and its instincts as a whole. The connexions

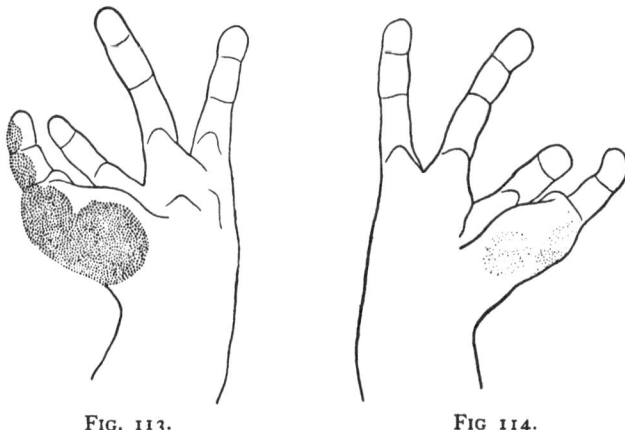

FIG. 113.   FIG 114.

FIG. 113.—Hand of a normal male frog, *Rana fusca* (*R. temporaria*).
FIG. 114.—Hand of a castrated male frog, *Rana fusca* (*R. temporaria*), about one year after castration. (After Meisenheimer.)

between the gonads and the secondary sexual characters are particularly striking, and are the easiest to investigate.

A case which is easily demonstrated is the dependence of the formation of the thumb pad of the male frog upon the condition of the testis. Normally this pad is developed only when the testis is present. After castration it either degenerates or, if castration is carried out between the breeding seasons, it does not attain its characteristic size and differentiation in the breeding season (Figs. 113 and 114). By subsequent implantation of pieces of testis, or by injection of testis substance into the lymph sacs of the castrated animal, the thumb pad again attains its normal form. At the same time the pad of the castrate is still more or less subject to its normal cyclical change of condition. If adult frogs are castrated the pad does not finally disappear, but at the proper season becomes somewhat more marked, as in normal animals. Now, even after implantation of ovarian substance the pad of the castrate begins to grow. This action, therefore, seems not to be one which is specific for the testis secretion, but a general influencing of metabolism through increased activity of the gonads in the breeding season. At the same time, the development of the thumb pad in the castrate after injection of ovarian substance is somewhat less pronounced than it is after implantation of testis.

In Birds the dependence of secondary characters upon the internal secretion of the gonads can be very clearly demonstrated, since a pronounced sexual dimorphism often exists in these animals. The relation of plumage to the gonad hormone has been most completely investigated. That such a relation exists follows from "age castration" in hens. In old hens the ovary degenerates progressively, and the plumage takes on a male character—"cock-feathered" hens thus arise. These observations are confirmed by castration experiments. In general, castrated hens develop cock-feathering, but the converse does not hold. The cocks of certain breeds of domestic fowls are, however, normally hen-feathered (e.g. Sebrights, Campines, Hamburgs). Castration of these cocks produces a typical cock-plumage. If after castration a gonad of the opposite sex (heterologous) be implanted, a type of plumage appears which corresponds to this gonad; so that hens with implanted testis acquire cock-plumage, and cocks with implanted ovary acquire hen-plumage. Castrated ducks develop typical drake-plumage: the characteristic changes in the plumage are the colour and form of the so-called "drake feathers" on the rump (Pl. XII, Figs. 115–118). These reactions are especially clear if young animals are used in the experiments.

Now, in Birds the special differentiation of the plumage is clearly not determined solely by the gonad hormone, the action of which is in the main one of realization. This appears from the facts that sex-linked factors play a part in the production

PLATE XII

FIG. 115

FIG. 116

FIG. 117

FIG. 118

FIGS. 115 to 118.—Drake-feathering of castrated ducks. Fig. 115, Rouen drake; Fig. 116, Rouen duck. Figs. 117 and 118, drake-feathered ducks of the same breed. (After Goodale.)

PLATE XIII

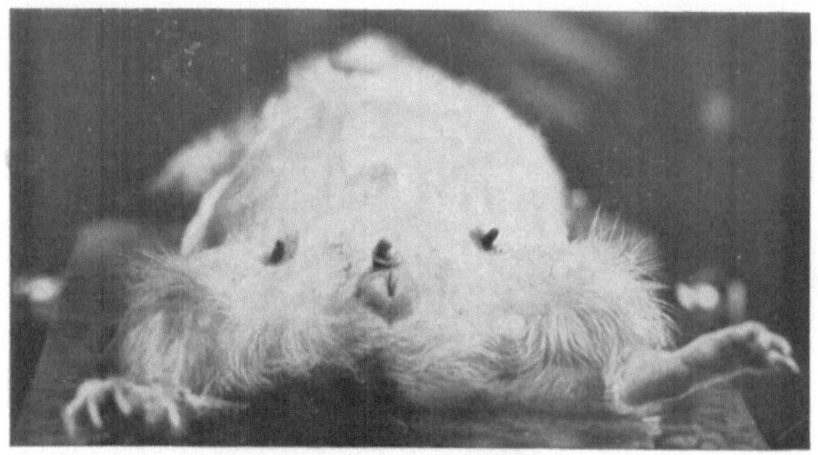

FIG. 119.—Teats of a feminized male guinea-pig in the period of lactation.

of the plumage, and that in unilateral gynandromorphs one side is feathered as a male, the other as a female, according to the side in which the corresponding gonad lies. Such unilateral gynandromorphs are not uncommon in finches. They possess both kinds of gonads, the ovary generally being on the left, and the testis on the right. Corresponding with this, the left half of the animal exhibits female plumage; the right half, male. The division between the two kinds of plumage exactly coincides with the median plane. In such cases the explanation of the secondary sexual characters cannot be based on hormonic action by the gonad, since hormones, diffusing as they do throughout the body, must be effective in every part of it, no matter on what side the source of the hormone lies. An explanation must therefore be sought in factors other than hormones.

In Mammals certainly very many, if not all, of the secondary sexual characters depend for their complete formation on the gonad hormone. This is proved by numerous pathological observations and by many experiments. Quite generally after castration the form of male mammals, including man, assumes a female character, especially when the gonad is removed early.

As a particular example we may mention the well-known dependence of antler-formation in the Cervidæ upon the condition of the gonads. Absence of the testes, or total castration of the young animal, suppresses the development of antlers, and moreover gives a female character to the skull. If atrophy of the testes occurs later in life, the antlers form a "peruke." This is most frequently seen in the roe-deer, but it may occur in any of the Cervidæ. What happens is that the "velvet"—the skin covering the young antlers—does not wither, and consequently is not thrown off; the bone of the antler begins to proliferate irregularly, and in this way there may be produced what looks like a large fur cap. Conversely, damage to the ovaries of females, which normally are without antlers, leads to their formation.

Now, the relation of the antlers to the testis is complicated by the fact that it is clearly not purely hormonic in nature. For it is stated that innate unilateral atrophy of a testis produces a stunting of the antler of the opposite side. Abnormal formation of the antlers may also arise after injuries to the soft parts and bones of the limbs, and to the pectoral and pelvic girdles. There is present, therefore, a complex of different kinds of relations.

The bearing of the gonads on the appearance of the secondary sexual characters and on the whole sexual *facies* can be demonstrated by implantation of heterologous gonads after castration. If young male rats or guinea-pigs are castrated, and ovaries are then implanted in them, the ovaries remain fairly healthy for some time, and the influence of the ovarian hormone becomes apparent. The implanted ovary does, indeed, finally break down to form interstitial tissue, but even then its influence on the somatic characters is maintained. Castrates with an implanted ovary become feminized; mammary glands and teats are developed identical in size and shape with those of the female (Pl. XIII, Fig. 119). Moreover the general appearance, and the skeleton, of the feminized males become like those of a female; conversely females which have had a testis implanted into them, after having been castrated, acquire male characteristics. The normal sexual impulse is absent in males which have been feminized; they shew periodic enlargement of the mammary glands, periodic secretion of milk, and the capacity to look after and to suckle the young. A corresponding alteration of the whole behaviour takes place in masculinized females.

The much discussed question of the origin of the gonad hormone has not been finally decided. According to one view, this secretion is a product of the interstitial cells; according to another, only the actual generative cells are able to produce it. It is quite possible that both may be concerned.

From the point of view of its general effect the gonad hormone is the most important of all internal secretions. More than any other of these it influences all the characters of the body, besides playing an important part in the maintenance of its typical condition. The possibility presents itself here that disappearance of the normal characters—i.e. the phenomena of old age—might be combated by this hormone. In other words, a kind of rejuvenescence might be produced.

Experiments designed to test this must aim at augmenting the hormone at a time when the internal secretions of the gonads are reduced in quantity, and, if possible, at reviving the power of internal secretion in the glands themselves. In male mammals this can be done in two ways: either by ligaturing the vas deferens of one or of both testes, or by implanting a young gonad in the body. In the first case regenerative processes are set up in the testis, and at the same time there is an increase in its internal secretion. In the second case the implanted gonad first acts as a source of the hormone, and this in its turn favourably affects the indigenous testis. Only the implantation of the young gonad is possible in female animals.

Both methods have, in fact, been successful. Senile rats, after the testis has been isolated by ligature, become lively again in their behaviour; appetite and weight increase; bald patches on the skin shew a new growth of hair, and the fur again becomes thick and smooth; the sexual impulse reawakens, and the animals are able to breed. In other mammals similar experiments have been successfully performed. This operation is of practical importance in the case of valuable breeding stock. Rejuvenation in man has been attempted in this way, but opinions about it are still divided. There have been failures but fairly numerous positive results have been obtained. This is also true of the transplantation method. Sometimes good results have followed the implantation of a young ovary into female animals. This implanted ovary serves simply as a source of the hormone, and is not a substitute for the senile ovary of the "host" as a formative organ for the germ-cells. These last are formed by the reactivated ovaries of the host, and it does not matter therefore in what position the implanted ovary is placed. By implanting under the skin of the throat ovaries, or parts of ovaries, from castrated or slaughtered animals it was possible to reactivate a large number of pre-senile cows, so that they became pregnant and calved. It must, however, be remarked that the widespread notion that rejuvenation can be easily produced by such operations should be accepted with reserve.

(c) Interrelations due to physico-chemical agency

Now, in dependent relations, factors whose action is directly physico-chemical become important. This is specially true of the interference of external factors with the process of realization. It has been pointed out in an earlier section that the concentration of substances in the medium may influence the course of development. We will remind the reader again of the action of common salt upon gastrulation in the Amphibia, where a certain amount of this or other salts in the culture-water delays or completely inhibits the process. Invagination of the vegetative region is held up, and a giant yolk-plug is formed— sometimes an exo-gastrula. It is not too much to assume that this is not the result of a specific action of the salt, but simply that the viscosity or the condition as regards swelling of the plasma-colloid of the germ is so affected by the altered osmotic conditions that its viscosity increases, thus hindering the formative movements. The results of these purely mechanical actions of the medium may become so complicated as indirectly to involve relations which are conditioned by stimuli; but the real action of the environmental factor is here simply a direct physico-chemical one. Among many other phenomena the falling apart of blastomeres in calcium-free sea-water is in the same category. On the other hand, however, the relation of external factors to development is not always by way of directly mechanical action. One need only recall the importance of light in the formation of pigment in the pupæ of *Pieris*, to recognize that external factors may interfere also as stimuli in the process of realization. This holds for many other cases of the action of external factors.

In general it is true to say that in the manifold relations present in development there exist a co-operation and an interplay of the most diverse kinds of activity; and though purely physico-chemical agency is not concerned in truly determinative interrelations, there is certainly, in the action of stimuli, an extraordinary complexity which is not yet well understood. In the great group of realizing relations too, the existence of extremely complicated connexions must be assumed; but the fact that definitely individualized activities can be recognized in a particular process must not lead us to conclude that development as a whole is nothing more than the sum of these relatively simple physico-chemical actions.

III. Independence of the Parts

If, on the one hand, development is dominated by numerous and complicated interrelations and dependencies, we have at the same time to admit that parts may shew great independence of one another in development. This is not surprising; because in the case, for example, of determinative relations, these relations become unnecessary or inoperative as soon as determination is completed. The same is true of most of the realizing relations; these again are really indispensable only until their action has begun to take effect. Thus the developmental independence of the parts does not contradict what has previously been strongly emphasized—the presence of an interlocking of all the parts and process in development. In normal development dependence and independence alternate with one another. Independence is particularly striking in experiments where the parts of different early embryos are artificially united. If the parts in question are already determined, their development goes on independently—from the nature of the case no reciprocal influence is possible. In heteroplastic combinations particularly, the specific character of the one partner is never affected by the other, so that even when development gives an integral product the heterogeneous origin of the components remains apparent. This is true of chimæras. Sectorial and periclinal chimæras can be distinguished. The former arise if two blastomeres of different species are joined together to form an artificial whole germ, or if from two half-gastrulæ of different species a new whole gastrula is formed. Although in the latter case the right and left halves of the body have arisen from different material, a morphological and physiological whole can nevertheless be formed, without the original specific character of the halves being lost. Periclinal chimæras appear when, as the result of the junction of regeneration-blastemas, a limb-regenerate is formed, the skin of which belongs to the one species, the tissues to the other. It has not been possible hitherto to manufacture whole animals which are periclinal chimæras, as has been done in the case of sectorial chimæras; but such whole periclinal chimæras have been observed among plants.

CHAPTER VIII

THE GERM-CELL AS REACTION-BASIS

I. The Reaction-Basis from the Point of View of Genetics

1. The Reaction-Basis and the Theory of Factors

The conception of the reaction-basis has already been explained (p. 121). The reaction-basis is, in the last analysis, the real foundation of the internal factors of development, and is expressed in the specificity of the germ-cells. The problem of its actual nature can be attacked in various ways, among which experimental analysis of the mode of action of internal factors and of the role of individual germ-regions in determination must take the first place. But every other line of investigation which concerns itself with the internal factors, or with the analytical study of the germ-cells, contributes to the solution of the problem.

Genetics throws many side-lights upon the nature of the reaction-basis. What is studied in genetics is really the behaviour of the internal factors of development, not indeed from the causal point of view of their mode of action, but with reference to the form assumed by their end-products in the sequence of generations, and to the distribution of the internal factors among the descendants. Though that be so, it can indirectly throw light on the problem of the reaction-basis.

The first concern of genetics is not the reaction-basis. On the contrary, its starting-point is the behaviour of the external characters in the sequence of generations, and in particular their statistical distribution. Since, however, external characters (in so far as they are not modifications) are the product of internal developmental factors carried by the reaction-basis, we must, in the end, attempt to relate these characters to peculiarities of the reaction-basis. Though from the statistically determined frequency and distribution of external characters it is possible to draw conclusions about the behaviour of the internal factors of development—genes, as they are called in genetics—a picture is in this way obtained not of the reaction-basis but of the genotype—that is, of an aggregate of all the genes. The reaction-basis is the actual bearer of this genotypic peculiarity of the individual or of the race; but it can only be understood by the aid of other methods of investigation.

Modern experimental genetics is a younger science than experimental embryology, but from the first it has taken its own course independent of the latter. Genetics sought—and found—its complement in cytology, and the result has been that, in genetics, views concerning the nature of the reaction-basis have been almost exclusively limited by what is a purely morphological field of investigation. Notions concerning the nature and special characters of the internal factors of development and of their carrier, the reaction-basis, lead naturally to definite conceptions of the nature of development; and since the function of development is the most specific of all the phenomena of life, they lead also to definite conceptions of the organism and of life in general. Now experimental embryology too has developed, as a result of its own findings, certain fundamental conceptions, which have already been stated in this book. We must now call attention to the interesting fact— the importance of which should not be underestimated—that the fundamental conceptions of genetics relating to the nature of development and of the organism stand in sharp opposition to those of analytical embryology.

Naturally a synthesis of the two points of view of the sister sciences is possible; it is equally obvious that the greater weight should not be given to that science whose starting-point is the end-result of the process of development—the realized external character—but to that science which is concerned with the actual basis of development, and with the process by which the final product is attained—the science of experimental embryology. Only in this way can the nature of development be understood. In so far as the views of genetics are antagonistic they must undergo modification.

The ideas which prevail in the science of genetics start from the assumption—one might call it the axiom of modern genetics —that inheritance is due to independent hereditary units, generally called hereditary factors or genes. This fundamental principle has undergone many transformations and expansions with the progress of knowledge, which has called for additional hypotheses; but it remains, as at first, the real foundation of the whole edifice of the science, and particularly of the factor theory. No opinion is originally implicit in this principle as to the concrete nature of the hereditary units; there is only assumed something active, some factor; its localization in the germ-cell and the question of its possible material basis do not concern the theory of factors. Moreover, these things cannot be decided solely by a consideration of statistically ascertained phenomena of inheritance. We must rely on other observations.

The simple principle of the hereditary unit makes it possible to explain many of the phenomena of inheritance. From the combination and separation of hereditary units which occur during fertilization and gametogenesis, and which are governed by chance (or, better, by probability), the rules according to which definite characters appear, disappear, and reappear in inheritance can easily be deduced. The content of the law of Mendelian segregation, the phenomena of dominance and recessiveness, and the intermediate behaviour of hereditary factors are also quite in accord with this fundamental principle.

Now it is at once apparent that certain occurrences in the process of inheritance are not in agreement with the assumption of completely independent genes; frequently it happens that several or even a great number of characters are always inherited together, so that the factors responsible for them cannot, in the sequence of generations, be other than combined with one another. This leads to the conception of the *linkage group*, by which is meant a combination of a number of genes, which as a rule is not broken up but is passed on intact from one generation to the next. In certain cases a gene may, indeed, leave such a combination to become temporarily "free"—an assumption necessary in order to explain the occasional free "mendelizing" of a character ordinarily coupled with others. We speak here of a "crossing over" of factors. This can only be seen in hybridization, for in that case the gene which becomes free is exchanged for a contrasting homologous gene, so that in the descendants a character, itself belonging to the linkage-group, is replaced by another which originally formed part of a linkage-group of the other race, and vice versa.

Generally each zygote contains every linkage-group paired—one member of the pair from the male, the other from the female. Only one particular linkage-group occurs unpaired: the one containing the sex-linked factors. The inheritance of sex as a back-cross of a heterozygous with a homozygous individual renders necessary this assumption, which in any case is in agreement with the observed behaviour of sex-linked characters in inheritance. In Mammalia, Diptera, Orthoptera, and others this unpaired linkage-group appears in the male; in Lepidoptera, Birds, and some other kinds of animals, in the female.

2. The Reaction-Basis and the Chromosome Theory

According to current ideas the material basis of the linkage-groups, and therefore also of the individual genes, is to be found in the chromosomes, which—in the phraseology here adopted—would therefore have to be regarded as the real reaction-basis. This conception rests chiefly on the following observations and lines of argument.

If definite (formed) elements of the germ-cell are conceived to be the material carriers of the "mendelizing" genes and of the linkage-groups, they must fulfil a number of conditions laid down by the factor theory of inheritance, as follows.

The structures in question must be present in the cells of male and female individuals in equal numbers, and this number must be constant in all generations. In view of the appearance of the linkage-groups in pairs in the zygote, and therefore in the cells of the individual, we are limited to structures that appear paired in the fertilized egg and in all the cells of the body. In the ripe reproductive cells, however, each kind of "carrier" must only be present unpaired, for if not there would be a rapid increase in their numbers. The unripe germ-cells, on the other hand, must shew the full double number of the corpuscles in question, and the maturation of these cells must involve the division of each pair of corpuscles into its two components, so that only one of each kind of these components finds its way into one and the same ripe germ-cell. The factor theory of sex-inheritance and sex-linked inheritance demands that those factor-carriers which contain the sex factor, and the somatic factors coupled with it, should also, in certain cases, appear unpaired. Further, among Mammalia, Diptera, and Orthoptera, they must be present paired in those fertilized eggs which produce females, unpaired in those zygotes which produce males. For the other type of sex-inheritance, found in butterflies and birds, the reciprocal behaviour must hold: we ought to find in the female zygote an unpaired corpuscle, in the male paired corpuscles as carriers of the linkage-group. The crossing-over of factors must also be in harmony with the intracellular structures assumed to exist.

Now, it is clear from a study of the maturation of the germ-cells and of the process of fertilization that it is the chromosomes whose behaviour fulfils these conditions. The number and constancy of chromosomes at their appearance in cell-division, maturation, and fertilization; the existence of an unpaired (X-) chromosome which has been demonstrated in many, if not in all, of those zygotes where an unpaired linkage-group must be assumed,—these are not the sole considerations which point to the chromosomes as the carriers of the linkage-groups. A further and more weighty reason for this opinion is to be found in the fact that, in certain cases, the number of linkage-groups assumed to be present on the factor hypothesis coincides with the number of chromosome pairs. This is true for the Fruit-fly *Drosophila melanogaster*, in which four linkage-groups and four pairs of chromosomes are found. Again, these linkage-groups obviously vary in extent, and parallel with this variation there is a difference of size in the chromosomes. The "mendelizing" of the chromosomes—that is, their distribution according to probability during the maturation divisions—has been demonstrated by the investigation of hybrids, in which the members of the pairs of chromosomes are unequal in size, since the chromosomes of the races crossed are unequal in size. It is thus possible to locate the chromosomes during fertilization and maturation, and again in the subsequent fertilization, and to shew that the positions they take up correspond to the expectation on a random distribution.

According to the ideas current in genetics, the chromosomes are therefore regarded as the material bearers of inheritance— indeed, as the sole bearers. Hence heredity in a linkage-group simply means *heredity in one and the same chromosome*. Each chromosome is a combination of hereditary units or genes. And by gene is meant not simply, in a general way, an inherited factor of development, but something very definite—a material corpuscle situated in one definite part of one particular chromosome. Crossing-over corresponds with an exchange of pieces of chromosomes, which contain, according to their sizes, more or fewer of the corpuscular genes. There exist no carriers of inheritance—no determiners of development—apart from these chromosomal genes. For every essential factor-symbol in the genetic formula of a race there is, in fact, a material particle of this kind in a definite chromosome. An attempt has even been made to assess the exact localization of the individual genes within the chromosome, and to make in this way so-called *chromosome maps*, in which the chromosomes are represented as lines with the individual genes as definitely localized points lying upon them.

Out of the above elementary principles (with certain additions which need not be referred to here) the science of genetics has built up a theory which, for its own purposes, is extraordinarily serviceable. Starting with the phenomena of heredity, and having as its goal the analysis and representation of the process of inheritance, this theory completely achieves in its own field what it sets out to do. There is no necessity to replace it, nor is it easy to see at the moment how this would be possible. There may be certain phenomena which do not easily fit into this theory, but they can generally be made to do so by the aid of relatively insignificant additional hypotheses which are in accord with the basis of the theory as a whole. Geneticists are entirely in the right when they regard the factor theory and the chromosome theory of heredity as truths which approach finality. It must not, however, be forgotten that this is a theory which has been advanced to cover a definite, limited field of phenomena. As soon as we go beyond its limits certain difficulties are met with which lead to a conflict with the fundamental ideas of experimental embryology.

The present position is that the chromosome theory of heredity is undoubtedly extending its bounds so as to embrace a more comprehensive theory of development. For from it are deduced definite views about the nature of the reaction-basis, which views are no longer generally regarded as theoretical formulations, by which we explain definite processes in our description of heredity, but as descriptions of actual, concrete things.

Seen in this way, the reaction-basis appears as a mosaic of material corpuscles. Genetics starts with the study of special (accessory) characters, so that the factor-formulæ of the genes primarily refer only to these characters. It is doubtful, however, whether the special characters and the essential characters of the organism can be clearly separated from one another. Disregarding this, ideas concerning the inheritance of special characters are usually extended to cover the behaviour of general or type-characters. For these latter also the reaction-basis must, in that case, contain the appropriate corpuscular genes. These genes are parts of individual chromosomes, and it is supposed that for each essential and each accessory character there exist one or more discrete material particles in a definite position in one definite chromosome—or possibly in several. The essential characters determine the typical structure of the organism, i.e. its organs and everything belonging to them— its specific structure as a whole, which places it in the type to which it belongs. The accessory characters are peculiarities of these organs and of this structure, which can exist within the type in large numbers without altering its nature. But, for all these, discrete corpuscles are present in the individual chromosomes of the germ-cells from the beginning.

3. Deductions concerning the Nature of Development and of the Organism

The importance, from the point of view of analytical embryology, to be ascribed to this material mosaic, forming the foundation of every organism, is directly apparent when one takes into account the way in which the conception has arisen. There are associated with the individual heritable characters definite causal factors, which produce them; these factors—on the basis of the theory of factors and the chromosome theory—are pushed right back to the beginning of development, and are regarded as concrete corpuscles in the individual chromosomes. In other words, these primordial corpuscles have a direct relation to the essential or accessory character produced by them. To follow this line of thought consistently, they must be regarded as the determiners of development; and it is not surprising, therefore, that the prevailing opinion in genetics— which is the science of these corpuscular genes—does in fact accept them as the sole determiners of development. The remaining constituents of the cell, the cytoplasm in particular, are at most considered as being a substratum for the activities of the particles contained in the nucleus. Seen in this way the whole process of development is an Evolution pure and simple, since there is an actual preformation of every part and of every peculiarity: for the connexion between corpuscular primordium and realized character is throughout a direct one. In the concrete mosaic of the chromosomal reaction-basis, any one material rudiment represents its own single differentiation, whether this last be a particular organ or a particular character of an organ. All the variety of differentiation and morpho-genesis shewn at the close of development, even in the most complicated and specialized of organisms, is already actually and materially present in the chromosomal basis—not, indeed, as the old theory of "emboîtement" demanded, in the form of miniature organs enclosed in the germ-cells, but nevertheless materially present, because there are, from the beginning, at least as many discrete chromosomal particles as there are later essential and accessory characters. Since, however, in order to explain inheritance, it is necessary to assume that a number of corpuscular genes are often concerned in the production of a single character, the conception leads necessarily to our imagining a greater manifoldness at the beginning of development than at the end. But quite apart from this, on the supposition of any such basis of development, the whole developmental process can only consist in an unfolding of what already exists: it can only be a piece of mosaic-work, as it has indeed been well called. As the stones of a mosaic, themselves discrete parts, combine together to form a single picture, so the independent activities of the separate primordial corpuscles create, in the end, a whole organism by the simple self-differentiation of its parts. Everything in development is preformed and determined in advance by chromosomal corpuscles, the nature of which is extraordinarily constant.

Since the rudiments, collectively, are material in nature, their mode of action is mechanistic—i.e. mainly chemical and physical. There is no place for any other activity. The chromosomes are, to put it graphically, aggregates of microchemical laboratories, and development depends on pure mechanism even though we may avoid using the word machine.

Such a conception does not contain anything that is new in principle, and did not first arise from modern genetics. It was developed, though in a somewhat different form, and not so definitely in relation to particular parts of the cell, by Weismann years ago in his theory of the Germ-plasm. It seems, however, to have gained in prestige as a result of the collaboration, which we have mentioned, between genetic analysis and cytology.

In defining the character of development, we express at the same time the essential nature of the living organism itself. If development is brought about by a mechanism, then the organism also can only be such a mechanism in the physico-chemical sense. All the processes of life, including development, must without exception then be purely mechanistic.

But this is not the only outcome of this conception of the organism. There is another, the importance of which is no less far-reaching, which is inseparable from the whole conception, and that is the *analytical-summative* representation of the individual, according to which the first rudiment of the individual consists of a sum of discrete parts—the corpuscles of the chromosomal reaction-basis. Its unity is thus a secondary thing, arising only from the interaction of its primary parts. How the unity and individuality of the organism come about is not explained by the current theories; the question is left open.

The interpretation of the individual which arises from the chromosome theory leads naturally to a method of presentation according with the classical cell-theory. According to this the organism consists of cells—not simply in the sense that cells are present in the metazoan body, which is a fact, not a theory—but in the sense that the body of the organism is a cell-state made up, on the principle of the division of labour, of individuals—the cells—which are more or less independent. The essential thing here is that the body of the higher animals and plants, is a secondary unit, which in the strict sense of the words, is *built up from* primary elements, the cells. The idea that it is a secondary unit is based on both phylogenetic and ontogenetic grounds; phylogenetic, because of the view that the metazoa have arisen by the coming together (or remaining together) of independent unicellular organisms; ontogenetic, because of the idea that the egg, during cleavage, is simply cut up into similar, independent cell-individuals, and that by the secondary union of these an individual of a higher order is formed—i.e. the cell-state of the metazoan body. The functions, taken collectively, of such a secondary unit must be explained in terms of the functions of the primary units; morphologically and physiologically the single cell comes first; the complete individual is a sum of parts, its functions are the sum of the functions of these parts.

Now, to take another aspect, the cell-theory is older than the theories of analytical genetics, and has exercised an extraordinarily far-reaching influence upon biological thought. If we look at the facts without prejudice, we see not only that the analytical-summative conception of genetics leads to the ideas of the cell-theory but, conversely, that this conception has arisen from the older cell-theory, and leads us beyond the elementary units originally imagined. Thus, it must not be assumed that conclusions from genetics about the essential nature of the individual are a confirmation of the cell-theory; on the contrary, the fact really is that genetics has derived its summative-analytical mode of thought from the older morphology.

Morphological analysis has not been content merely to demonstrate the cellular constitution of the body, but with the progress of cytology the cell itself has been analysed. In so far as this analysis consists of a description of the internal differentiations of the cell, it forms a natural and necessary continuation of the morphological investigation of the organism. Just as in the cell-theory of the whole organism the relation of the elementary unit of form to the individual has been fitted into a definite system, something analogous has happened in the case of the constituent parts of the cell. Their relation to the whole cell has been defined, either by assuming that the cell itself is a secondary unit, composed of yet simpler vital elements, as in the case of the cell-state of the whole body; or, in the more modern way, by conceiving the cell (particularly, the germ-cell—and, to be still more exact, the reaction-basis) as being essentially a sum of separate corpuscles, as it is expressed in the extreme form of the chromosome theory of inheritance. In this way the individual is completely resolved into a mosaic of practically independent parts: it becomes a mere summation.

II. The Reaction-Basis in the Light of Analytical Embryology

1. The Theory of the Unity of the Organism

The general conclusion to be drawn from the results of analytical embryology as to the essential nature of development has been stated at some length in a previous section. Development was seen to be a real epigenesis, and not an "evolution" depending on preformation. This follows quite clearly from investigations into the potency of parts of the germ, and from the phenomena of regulation; it is further emphasized by the uncertainty of the developmental fate of the parts, which has so often been demonstrated. This demonstration applied not only to the beginning of development and the early stages of the germ, but also to regeneration. The organizing and inductive action of subordinate organizations of the germ can only be interpreted by assuming that development is epigenetic, and this is true also of the progressive organization and determination which create only step by step the definitive factors active in the separate processes of development.

For all these reasons we conclude that the relation of the rudiments originally present in the zygote to the differentiations in development is not a direct one, but is indirect in the sense in which we have previously used the word. Such an indirect relation, however, is only possible when the basis of development consists not in the presence of a preforming corpuscle for each differentiation, but in the general character and constitution of the initial state of the germ. Only in this way, too, can we explain the manifestations of potency after development has been disturbed, and the phenomena of regulation.

Further, development does not rest exclusively on the independent action of parts, and is not simply a summation of numerous single processes: it is a single performance in itself. The separate parts—certainly the intracellular parts—are indeed of great importance, but only as subordinate organizations of the whole; their determining action and their differentiation are accomplished only within the framework of the whole and with reference to the whole. It is unnecessary to enter into further details on this point. This absolute dependence of the formation of parts upon the whole is specially noticeable in the phenomenon of multiple activity, of which many examples have already been given.

As we have said, it is quite certain that the analytical summative conception of the reaction-basis is largely derived from the theory of the cell-state. Opinion tends more and more to substitute for this purely analytical theory another, the germ of which is some years old, and which may be called the Theory of the Unity of the Organism (*Einheitstheorie*). Its central idea is that the metazoan body, though made up of many parts, is a primary unit, and that the parts are secondary. The individual cells, again, are not primary but secondary. It may perhaps be put in this way: cells do not form the body—the body possesses cells. This interpretation is supported by very strong evidence.

The old cell-theory has three aspects—a morphological, a physiological, and an evolutionary. In the first aspect, it is by no means applicable to the body as a whole. For this when fully differentiated does not consist of cells alone; it contains, widely distributed, many structures of other kinds, which are certainly not negligible, nor are they merely by-products of the cells. The cell theory also cannot account for the arrangements subsisting between the cells, on the one hand, and the whole organ or body, on the other. The phylogenetic argument which is usually advanced to support it is without force, since it is based on pure speculation. In its physiological aspect the old theory regards the physical manifestations of life as forming a mosaic-work. But this contradicts our general experience that the body functions throughout as a unit. The tonic state of the neuro-muscular system as a whole, and the unity of consciousness, demonstrate that the body cannot, even in theory, be considered an aggregate.

From the point of view of evolution the cell-theory is correct only in asserting that development has as its starting-point cells from which the other morphological elements of the body arise, and that in later stages also the cells play an important part. It is wrong, however, to regard development of the body—in the idiom of the cell-theory—as being the result of the colonial union of independent organisms, the cells. The germ remains a unity even after cleavage. The evolutionary importance of the individual cell has been much overrated in the old theory; for it is not the separate embryonic cells which are responsible for development, but the whole germ which carries out development by means of cells.

As would naturally be expected, development in its beginnings is controlled by the fertilized egg, but in its later course the cell, as such, is seen to recede more and more into the background. It need only be pointed out that the rate of cleavage is of great importance, the direction of cleavage of quite secondary importance, in early morphogenesis. For example, in gastrulation we see the activity not of cells but of the whole germ: the individual cells follow passively the morphogenetic movements of the germ as a whole.

The germ-layers, when such appear in development, are only subordinate systems within the germ, it is true; but, at the same time, each of them forms an integral whole, in which the part played by the individual cell is not one of capital importance. That division into cells does not—as on the old cell-theory—build up the organism and produce differentiation by the fitting together of a mosaic, appears also from the fact that there is no essential difference between regulative and mosaic eggs. Development in the latter is obviously independent of the formation of cells, although the secondary subdivision of the organism into cells is of great importance for the complete specialization of the various regions of the individual; for in mosaic eggs considerable differentiation has already taken place before the division into blastomeres. Regulative eggs appear somewhat different, simply because in them the point in time at which determination and differentiation take place is different from what it is in mosaic eggs; but in this matter there are, again, all possible transition stages. It has been emphasized already that cell-division does not itself bring about determination, and that the subdivision of the originally unicellular germ into many cells does not abolish the unity of the individual.

Epigenesis, which is the essential character of development, cannot be conceived as being an aggregate of separate processes connected simply with the individual cells; the problem of form can be grasped only from the standpoint of the Unity Theory. Processes of regulation, again, cannot conceivably take place in a partial embryo or a damaged embryo by the separate cells carrying out their proper functions as relatively independent members of a cell-state; these processes must owe their existence to the constitution of the developmental substratum as a whole, without reference to its subdivision into cells.

Development is, then, the affair of the whole organism; the individual cell is a secondary thing; and the germ is—and remains—the primary unit. We must not say, therefore, that the body during development is built up of cells, in the sense of the old cell-theory, but that it is subdivided into cells. The metazoan body is not a cell-state composed of almost identical and independent entities which have come together secondarily: its wholeness and individuality are primary, its parts and its cells are secondary.

2. The Unity of the Reaction-Basis

These considerations demonstrate the impossibility of extending the historical cell-theory: that is to say, of applying to the single cell—and particularly the germ-cell—its purely analytical-summative valuation of the individual. For the zygote is but the relatively undifferentiated unicellular state of the individual; the individual itself is a unity throughout its whole existence, even when it is a single cell.

If the reaction-basis be conceived as nothing more than a mosaic of discrete parts, then it is necessary to provide an "assembling" factor. The conclusion is inevitable that there is a Final Cause, an *entelechy* inherent in every organism—indeed in every individual cell—and controlling the mechanical system. From the standpoint of the Unity Theory such a dualistic view of the organism is superfluous; for the unity and entirety of the germ and of the individual are assured by the specific constitution of the cell as a whole. It is also certain that it is not the special properties of separate corpuscles, morphologically distinguishable from one another, that are alone to be regarded as factors of inheritance and development. The constitutive properties of the living protoplasm as a whole, which cannot be separated topographically, are factors too.

The general conclusion is, then, that it is not a mosaic of discrete corpuscles in the nucleus which forms the reaction-basis of development, but the whole germ-cell; and that sometimes one and sometimes another of the regions and subdivisions of the cell may assume the greater importance, according to the special character of the process concerned.

That which is inherited is a specific reaction-norm as a function of the whole cell; this last, in its entirety, is to be regarded as the reaction-basis, though there be a definite division of labour within its single individuality. If the germ-cell, as reaction-basis, does not consist simply of an aggregate of microchemical laboratories but, at the beginning of development, is a whole organism, the path is cleared for a more physiological conception of inheritance—the theories current in genetics being entirely morphological in character. The basis of development is the constitutive quality of the cell, including the subordinate organizations represented by its special structures and parts. Only in this way can the unity of what happens in development be understood, only so are regulation, potency of the parts, multiple assurance explicable. They cannot be explained on the basis of corpuscular rudiments forming a mosaic in which all the single parts are preformed. What is the same in all parts of the germ is its constitutive specificity. This it is which ultimately directs development, which makes possible the manifestations of potency, and the phenomena of regulation, and explains reactions to inductive influences in germ-regions not normally subject to them. None of these things would be possible on the basis of a preforming corpuscular mosaic.

Now the specific constitutive quality of the reaction-basis is not morphologically apparent, but the morphological differentiations of the germ-cell are not on that account without importance. So far as they are living constituents of the cell, these outwardly recognizable structures are to be regarded as the carriers of developmental factors. They are mechanisms of inheritance and development, and may be of great importance in individual processes. Among them are those constituents of the nucleus which make a transitory appearance in the form of the typical chromosomes. From the standpoint of analytical embryology, however, such mechanisms cannot be considered competent to act as the sole medium of the reaction-basis. There may, indeed, be cases in which mechanisms of this sort are apparently alone responsible for development, and therefore seem to be its sole determiners. But even here they can sometimes be set aside, and determination be completed without them, or even in opposition to them. Certain such cases will be dealt with later. We have, indeed, already encountered this phenomenon in the form of multiple assurance, the very possibility of which depends on the fact that every differentiation is not solely the result of one isolated mechanism but has its roots in the cell, in the germ, or in a portion of the germ—and always in the entirety of these.

Development starts out, then, not from an aggregate of separate parts, but from an integral reaction-basis composed of all those parts of the cell which are endowed with specificity. The zygote is not a summation of parts, but an individual organism in the unicellular condition.

III. RECONCILIATION OF THE OPPOSING THEORIES

The contradiction between the fundamental conceptions of genetics and those of analytical embryology is clear from the foregoing statements. On the one hand there is the analytical-summative conception that the organism is derived from a preforming mosaic of rudiments; on the other hand, the unitary conception of the organism as taking its origin from an integral reaction-basis functioning epigenetically.

It has already been suggested that it might easily be possible to obtain a unified conception by the combination of these two theories. The matter may be approached along three lines: first, the chromosomes must be given their proper importance in the cell as a whole; second, the epigenetic character of development must be borne in mind; third, the principle of progressive organization leading from the general to the particular must be taken into account. In attempting this unification our method must in part be a negative, critical one; in part, positive and synthetic.

1. The Question of the Concrete Existence of the Gene

(a) The continuity of the chromosomes

In heredity and development the gamete-nuclei and the nucleus of the zygote (and with these the chromosomes) are plainly of great importance; but the nucleus can also be shewn to be important from the point of view of experimental embryology. Two things can be regarded as certain: first, that the chromosomes play an important part in the production of linkage-phenomena; second, that where heterosomes (X- and Y-chromosomes) are present, normal sex-determination is usually brought about chiefly by the action of this chromosome-apparatus, which differs according to the sex. It is another question, however, whether the chromosomes play this part by virtue of being a mosaic of discrete individual genes, and whether they are, indeed, solely responsible for determination. It is quite possible, notwithstanding the importance of the chromosomes, that the truth is far different from this.

At the outset it must be kept clearly in mind that it is not actually the chromosomes—bodies appearing during the mitotic division of the nucleus—which form the source of the nuclear activity, but that this activity arises from the nucleus in the "resting" state. Not until division is complete, and the daughter nuclei have been established, does the real work of the nucleus begin. It is therefore incorrect to speak of a nucleus which is not in the act of division as a "resting" nucleus; it is better and more exact to call it an active nucleus.

If we develop logically the extreme chromosome theory, we have to imagine that the chromosomes exist unaltered in this active nucleus, and that during nuclear division they emerge from their association in the nucleus in the condition in which they always exist. For this alone could ensure the preservation, and the handing on from generation to generation, of the mosaic of rudiments with its normal arrangement of individual corpuscles.

The theory, often put forward, of such a continuity of the unchanged chromosomes even in the active (resting) nucleus arises from the conception of the individuality of the chromosomes. This conception is that the individual chromosomes in any given cell are qualitatively different: that they possess an individuality which remains unaltered in the sequence of cell-divisions. This idea is supported by good evidence, including the morphological behaviour of the chromosomes. The persistence of individuality must not on that account be taken to mean that the chromosomes in the active nucleus retain unchanged the relations and constitution that they always exhibit during mitosis. The great variety of changes undergone by the nuclear constituents during the life of the cell, both before and after mitosis, and which can be demonstrated cytologically, can only convey to the unprejudiced observer the impossibility of such a simple inclusion of completely unchanged chromosomes in the active nucleus. We can only be sure that the separate chromosomes appear again at each nuclear division in homologous individuality, and indeed in homologous morphological constitution, and that there is a continuity of their basis through all the generations of chromosomes and of nuclei. This, however, is not because they have remained unchanged in the active nucleus, but because there are definite material parts of this nucleus which are to be regarded as the foundations of the chromosomes that appear in mitosis, without being in every particular qualitatively identical with them. We may say that these foundations are homologous with the chromosomes, in the same way in which the swim-bladder and lung of vertebrate animals are said to be homologous—that is, without supposing for them a complete identity of nature and of functional importance any more than we do for the homologous organs cited as examples.

"Individuality" at mitosis is fundamentally different from "individuality" in the active (resting) nucleus. If this were not the case, it should be possible to demonstrate cytologically that the chromosomes in the active nucleus possessed exactly the same character as in mitosis—but this is not so. If the theory of the individuality of the chromosomes is transformed into the conception that they are permanently unchangeable and that their "mitosis individuality" persists in the active nucleus, we have an interpretation which follows the extreme chromo-some theory of inheritance, but which does not agree with the facts. It is no more surprising to find that in successive divisions of the nucleus the chromosomes always appear in the same number, form, and individuality, without being already present in the nucleus, than it is to find that the differentiated organism arises from the zygote without the individual parts of the former being directly enclosed in the latter. We find here, then, nothing that compels us to assume the existence of a persisting mosaic of corpuscular rudiments.

(b) The basis of the individual genes

In attempting a synthesis of the fundamental principles of analytical embryology on the one hand and of genetics on the other, it is very important to settle the question as to whether the corpuscular genes assumed in genetics do, or do not, in fact exist as discrete, material parts of the chromosomes. It is often taken as proved that they are actually present in the chromo-somes, definitely arranged in linear series; but a critical examination shews that not only has this not been established, but that the grounds on which the opinion is based are insecure.

Genetic research starts out from the statistical treatment of the incidence of definite external characters, and these are always non-essential or accessory characters by which particular races are distinguished from one another. In the factor-theory every separate character is based on one factor or more, generally designated by a letter or letters of the alphabet in the genotypic formula. How many of these factor-symbols, and which of them, are needed in this formula depends solely upon what symbols are necessary to describe the process of inheritance and to account for earlier and later generations. Factor-symbols are therefore introduced into the genotype formula as required; in the same way, according to the requirements of any particular case, either a simple or a complex factor may be used to represent the character in question. Hence the genotype formula is nothing more than an extremely exact description of the outward form of the process of inheritance, using a quite special mode of formulation which is excellently adapted to the purposes of genetics. In any case, the existence of the individual factors, so conceived, is at most a fiction designed to serve a definite purpose. It would be incorrect to suppose that because a new factor-symbol must be introduced into the genetic formula to describe the distribution of characters in the sequence of generations, that therefore the real, objective existence, as a discrete gene, of each single symbolized factor were proved.

If all the characters of the organism are due to such factors, the notions we have formed of the behaviour of the accessory characters must be generalized and applied to the essential characters of the organism also. These latter—that is the characters determining the type—must then be referred to such genes. Any generalization of this sort is in the highest degree open to criticism, for we know practically nothing about the inheritance process of the type itself; we know only that it arises again and again owing to the reaction-basis contained in the zygote. In any case, the fact that it is possible to describe the inheritance of accessory characters by means of a formula composed of the symbols of the separate factors is not a sufficient reason for accepting as proven the inheritance of the essential, or type characters by means of such factors.

Now, in genetics, these supposed factors are placed at the beginning of the whole process of development and are regarded as corpuscular constituents of the chromosomes. In this way, not only a real but a concrete existence is ascribed to them, and the necessity for adopting a single factor-symbol in the genetic formula is taken as proving the actual existence of a corresponding corpuscle in a definite chromosome. It is clear, however, that no proof of this is provided by a projection of the theoretical factors on to the chromosomes, though such a transference of the factors to the chromosomes is of use for the purposes of genetics.

Cytological investigation, again, does not at all favour the point of view that the chromosomes are series of corpuscular genes. While there is cytological evidence for a connexion between complete linkage-groups and chromosomes, so far as individual genes are concerned such evidence is wanting. An attempt has been made to represent the single corpuscular genes by the so-called *chromioles*. These are small structures, generally described as granules, which are to be seen in the chromosomes principally in the prophases of nuclear division. The chromosomes have the appearance of a string of beads, the single granules being intensely basophil in staining, while the connecting thread remains unstained.

Against the view that these small structures are corpuscular genes there is, however, so much to be urged that it cannot be sustained. In the first place the numerical relations must be considered. On the one hand it is not proved that the chromioles are constant in number, as obviously they must be were they parts of an unchanging mosaic of rudiments. On the other hand, their number—at least in very many cases—is too small for them to represent the genes. But there is the further and decisive consideration that their very existence as special structures is doubtful. Many things suggest that they may be artifacts caused by fixation. It is known, for example, that the action of a fixing agent on a colloid is sometimes to form small granules or hollow spheres. But apart from all this, they may be simply local thickenings of the substance of the chromosome; and this would sufficiently explain their strong staining reaction. In favour of this is the fact that in the living chromosomes constrictions may appear; and they may exhibit fairly rapid changes of shape, and contractions which are also of interest in this connexion.

The fully-formed chromosomes are, further, somewhat simple in structure, and in polarized light appear isotropic. Taking into account all that is known, we can only affirm with certainty that a chromosome consists of a thick, gelatinous material, impregnated more or less equally throughout with thymonucleic acid. At the same time, it may be that the actual chromosome does not simply consist of one completely homogeneous substance, but that it is some sort of mixture. Observation alone produces no objective evidence that a chromosome is simply an aggregate of discrete corpuscles which may differ completely from one another. Even though, in the prophases and the regressive phases of the chromosome (that is to say, during its formation from the nuclear reticulum, and again, later, during the development from the chromosome of the parts of the nucleus corresponding to it), definite structures make their appearance, this does not prove the presence in the chromosome itself of numerous corpuscular genes. It proves only that important processes of transformation are at work, as we have previously insisted.

In view of all this, it cannot be said that individual corpuscular genes can be demonstrated. The genes are no more than an assumption, invaluable in theoretical genetics, and legitimate for the special purposes of that science. But this is not an admission of their concrete existence.

(c) *The chromosomes and the active nucleus ("resting" nucleus)*

The chromosomes certainly form the most important constituent of the active nucleus. This does not arise from their being bound together by the nuclear membrane into a close but superficial union, for the relations are really more complicated. The nucleus is not a mere capsule enclosing the unchanged chromosomes. The latter are certainly simpler in structure than the individual nuclear territories, for the nucleus shews, even morphologically, in comparison with the chromosomes, a high degree of differentiation, which causes its appearance to vary according to its general condition and its phase of activity. The most important parts of the active nucleus are therefore not the chromosomes themselves, but what is developed out of them. Hence it is not correct either to regard individual regions of the nucleus as being simply chromosomes or, on the other hand, to regard the chromosomes when they have appeared as being mechanically separated nuclear territories—even as it were in a "condensed" condition. It would be nearer the truth to regard the chromosomes as representing, indirectly, the rudiments of the most important nuclear constituents, which last develop by reason of these rudiments though the individual parts of the nucleus are not contained, as such, in the rudiments. The chromosomes possess a lesser degree of diversity than the nucleus which arises from them. Since the epigenetic character of embryonic development has been established, the assumption that epigenetic processes are concerned in the origin of the active nucleus should not be surprising.

There is, as we have already said, a continuity of the chromosomes through all the cell-generations. This is not to say that they themselves remain unchanged in the active nucleus, but that their basis persists through the phases of mitosis and of activity. During the preparation for division the chromosomes arise time after time, in the same number, shape, and character, from this basis, though indeed with additions from the transformed nuclear territories. The epigenetic character of all these processes renders it unnecessary that the chromosomes should take over all the nuclear constituents concerned, and incorporate them as morphologically distinguishable elements. The chromosomes need only be established again in such a condition as to enable them to carry, indirectly, the developmental basis of the new active nucleus. This interpretation makes it clear that they are, in fact, more simply differentiated and organized than the active nucleus arising from them.

In so far as the nucleus of the germ-cell, and the cell-nuclei of the germ, influence development, this influence arises from the active nucleus. It would therefore not be correct to relate these actions directly to the chromosomes. Now, in genetics, inheritance is generally said to be by way of the chromosomes; but this mode of expression must not be misunderstood. As carriers of genetic or developmental factors it is the actual nuclei of the gametes which are concerned, and not the chromosomes, which are not contained as such in the nuclei. At fertilization chromosomes arise from these nuclei, so that, in so far as developmental factors are transferred by these chromosomes, the combination of the separate foundations of the zygote nucleus is due to the chromosomes. That part of later development which is conditioned by nuclear action is then carried on by the active nucleus of the zygote and by the active nuclei of the embryonic cells.

The chromosomes again play an important part in the maturation divisions of the germ-cells. Since their distribution to the daughter cells—either to the egg or to the polar bodies— is governed by chance, they stand in close connexion with the appearance of the phenomena of segregation and the genetic behaviour of linkage-groups. The matter ought really therefore to be expressed thus: inheritance (the transmission of factors from parents to offspring), in so far as it is conditioned by the nuclei, is due to the nuclei of the gametes; segregation and combination of the factors are due to the chromosomes. To speak of inheritance as being due to the chromosomes is to employ a somewhat abbreviated mode of expression. This is useful in genetics, especially since the nature and number of the chromosomes arising from the nucleus often give a very good indication of its quality. But we must be careful not to form false ideas of the status of the chromosomes in the cell from this use of the words.

2. Are the Chromosomes Solely Responsible for Determination?

(a) Non-chromosomal sex-determination

It is important in connexion with the present question to decide whether the chromosomes (in reality, the parts of the active nucleus to which they give rise) are actually the sole carriers of the determining factors, as they would appear to be according to many statements; or if grounds exist for believing that determination may be due to other causes as well. If the chromosomes be not the only "carriers" or determiners, then the reaction-basis cannot be simply a mosaic of chromosomal corpuscles.

In the development of the chromosome theory, sex-factors in relation to the heterosomes have played a special part, as we have briefly pointed out already. Explanations of sex-determination in terms of a chromosomal mechanism are supported by many convincing facts; indeed, in cases where such a mechanism is present it must undoubtedly be held chiefly responsible, in normal circumstances, for the determination of sex. At the same time, very many equally convincing facts in connexion with sex-determination by no means support this exclusive role of the chromosomes: in fact, they directly contradict it. There are cases in which sex-determination occurs without the aid of a sex-differentiated chromosome apparatus.

In two of the commonest species of frogs, *Rana fusca* (*R. temporaria*) and *R. esculenta*, different local races are found, distinguishable by their behaviour with regard to sex-differentiation. In certain races males and females, with typical gonads, appear from the first in equal numbers; these are appropriately called "differentiated" races. In other races, however, the gonads of all the individuals begin to differentiate as if to produce ovaries, and only secondarily are some of them converted into testes (Pl. XIV, Fig. 120).

Such an occurrence cannot be made to accord with the exclusive action of a chromosome mechanism; for either all the individuals have the same chromosomal constitution, in which

PLATE XIV

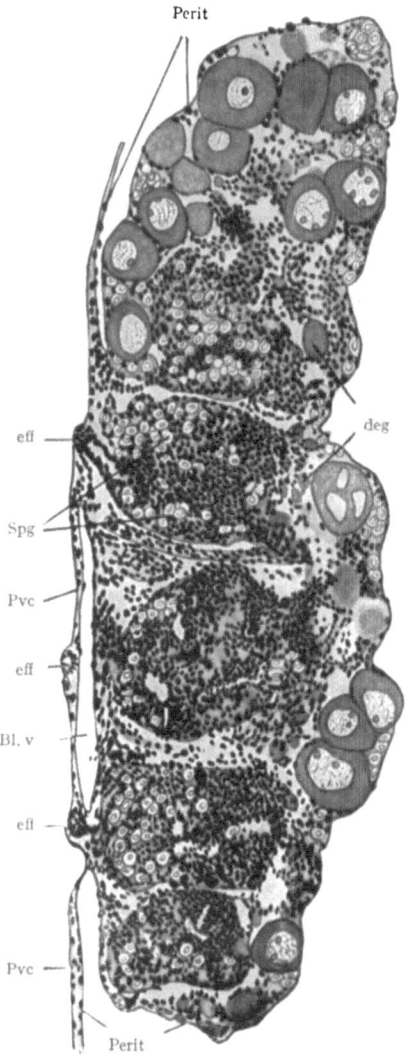

FIG. 120.—Gonad of a young frog during the change from its original female condition to that of a testis. At the periphery many oocytes are still seen. *deg*, degenerating elements; *eff*, vasa efferentia; *Bl.v*, blood vessel; *Perit*, peritoneum; *Spg*, spermatogonia; *Pvc*, wall of vena cava. (After Witschi.)

case it is impossible to understand the ultimate differentiation into two sexes; or they have different chromosomes, in which case the originally uniform female differentiation is unexplained. The only way out of the dilemma is to suppose that the chromosome mechanism is only one of the possible means of sex-determination. Such a mechanism is operative principally in the "differentiated" races, where males and females are produced in equal numbers from the beginning. In the other, so-called "undifferentiated," races the germ-cells develop into oocytes if they are in the germinal epithelium, that is if they remain in the outer layer of the gonad; they develop into spermatocytes, however, if they sink down into the gonad in the sex strands. Hence, all the animals are bipotential as regards sex, and it is not a chromosome mechanism which is decisive. Rather it is physiological conditions which are responsible—the difference in situation of the primordial germ-cells. Similar relations appear to exist in certain fishes.

A specially instructive case in this connexion is that of *Dinophilus*, a worm which produces two kinds of eggs—large female eggs and small male eggs. In spermatogenesis, spermatids of one kind are formed, each of which receives altogether 10 chromosomes (the haploid number). The ovary of the just-hatched female consists of a few oogonia, which multiply by normal nuclear division. After the first growth-period all the eggs are of the same size. Insemination takes place during the third growth-period—before the reserve materials of the egg have been formed, but after the differentiation into male and female eggs which involves sex-determination. Insemination and fertilization are therefore without influence upon sex-determination, and there is no question of alternative combinations of the chromosome apparatus being responsible. Closer investigation of the maturation divisions confirms this. Maturation is only completed in the egg after it is laid. Maturation and fertilization take place in all the eggs in exactly the same way. In the first maturation division 10 tetrads appear; 10 chromosomes remain in the egg; each fertilized egg has 20 chromosomes, the diploid number. No heterosome apparatus, differing according to the sex, can be demonstrated. The nuclei of the two kinds of eggs are distinguishable only by their size; any morphological cause for the sex-differentiation of the eggs is entirely lacking. Sex is already decided in the growing oocytes, i.e. in the unripe eggs; they have a male or a female constitution. Nothing definite can be said about the internal processes which here produce sex-determination, but we shall certainly not be far wrong in not referring this determination to a chromosome mechanism acting as the heterosome apparatus does when that is present.

In many forms of animals we see a cyclical change of sex. Generations in which either parthenogenetic females or hermaphrodites alone are found alternate with generations in which there is diœcious differentiation of the individuals. Some cases are encountered which accord well with the chromosome theory of sex-determination, while there are some also in which the chromosomes are not primarily concerned, but at most play a secondary part.

The Nematode worm, *Rhabditis nigrovenosa*, shews a change of this sort in its sex-differentiation. Its diœcious generation inhabits damp earth, and alternates with an hermaphrodite generation parasitic in the lung of the frog. In the diœcious generation, eggs of only one kind are produced, and these contain during maturation 6 chromosomes. Two kinds of spermatids, however, are formed, of which one kind—possessing 5 chromosomes—degenerates, while the other—with 6 chromosomes—produces ripe spermatozoa. From eggs fertilized by these arises the parasitic generation. Judged by their chromosome-number (12) they should be female; but they become hermaphrodite, some of their reproductive cells developing into eggs and the others into spermatozoa. The eggs have the same character as those of the diœcious generation, while the spermatozoa are of two kinds—one having 6 chromosomes, and the other, by reason of a peculiar reduction, 5 chromosomes. The unripe spermatogonia possess at first 12 chromosomes; thus, at the maturation division all the spermatids receive 6, but in half of these cells one chromosome always degenerates.

Fertilization with the latter produces males of the free-living generation (11 chromosomes), insemination with the other spermatozoa produces females (12 chromosomes). Thus in the case of the hermaphrodite, in spite of a female chromosome constitution, a male gonad arises; this argues strongly against the view that the chromosomes are solely responsible for sex-determination. Further, the elimination of one chromosome in half the spermatids cannot be without a cause. This may well be looked for first in the constitution of the cell as a whole, for to this the behaviour of the chromosomes is subject—here they do not govern, but are themselves governed. Secondarily, then, there is good agreement between the relations of the chromosomes and the chromosome theory, but what is primarily decisive in determination must be sought elsewhere.

In considering the source of sex-determination the change of sex of adult animals calls for special attention, because in such cases a perfectly definite chromosome constitution exists which, if it is of importance at all, can only be responsible for the original sex of the individual concerned.

An interesting case is that of the toad, *Bufo vulgaris.* All male toads possess, near each testis, a so-called Bidder's organ, which from its structure must be regarded as a rudimentary ovary. It is developed from the anterior part of the gonad-rudiment, the hinder end of which becomes the testis. If the testes are extirpated in adult males, and the castrates are then fed on an abundant diet, rich in fats (mealworms), they ultimately change completely into females. The caudal part of Bidder's organ becomes a normal ovary, its anterior portion a female Bidder's organ. From the rudimentary Wolffian duct, present in the male, are developed the oviducts; and the animals gradually acquire the complete female body-form, and assume the female reproductive functions, producing normal eggs capable of fertilization, and normal progeny. Since the chromosomal foundation of the nucleus has not been changed during this experiment, it is obviously a general change in metabolism which, in sexually bipotential animals, brings about the development of the female sex with all its primary and secondary organs.

In Birds also a change of sex in adult individuals has often been observed, as for example in the domestic fowl. A Buff Orpington hen 3½ years old suffered from a diseased ovary. Up to that time it had been a good layer, and had successfully hatched eggs. In the course of about two years it changed into a complete cock. This appeared not only from the plumage but from the behaviour of the animal, which was that of a normal cock—it mated with normal hens, from whose eggs were hatched male and female chickens, and these were inbred to produce typical descendants. Histological investigation of the gonad shewed the presence of two typical testes. Similar occurrences have been observed in other birds; and experimentally also it has been possible by early castration to transform female birds into true males.

Here, again, we are forced to the conclusion that if the original sex was determined at fertilization by a chromosomal mechanism, the determination of the secondary sex of the same individual cannot be due to the same apparatus. The chromosomes, therefore, cannot be at any rate the sole determiners of sex.

(b) The part played by the cytoplasm in determination

Now, just as it appears from the foregoing that we cannot regard the chromosomes as alone responsible for the processes of determination, it is possible to shew in the same way that the cytoplasm, for its part, is not simply the passive environment of the nucleus, but plays a very active part in development.

The cell is a physiological unit. From that alone one might infer that the cytoplasm of the zygote interferes actively in processes of determination, for the functional unity of the cell depends on a lively interaction between nucleus and cytoplasm; and, indeed, which of the component activities appears to predominate sometimes depends only upon the point of view of the observer.

In attempting to answer the questions that are here involved, the problem of localization must not be forgotten. It is not only necessary that the inherited factors should exercise a definite action on development and on differentiation, but that this action should take place at a definite point in the zygote and in the germ; for the individual is not a mere mixture of chemically different materials, but an organism with a typical internal and external structure. If the cytoplasm is simply an indifferent substratum for development, and the material particles of the chromosomes the sole source of the activity present in development, then the typical localization of activities in development remains completely incomprehensible. According to current theories, the idioplasm exerts an action which is purely chemical; such an action, however, can only be definitely localized in the cytoplasm if the latter, far from being an indifferent and homogeneous substratum, reacts in one place to some particular influence proceeding from the chromosomes, and in other places to other such influences. Or we may put it that the place where a particular reaction shall occur depends upon the collaboration of the cytoplasm. No theory of development, therefore, can merely leave the cytoplasm out of account. Local variations in the mode of action of the cytoplasm need not, of course, manifest themselves morphologically as differences of perceptible structure.

More convincing, however, than general considerations of this sort are certain observations, some of which have already been mentioned in another connexion.

In the morphogenetic processes of early development it is not the single cells which are responsible, it is the germ as a whole which carries out the movements involved. This is simply to say that the whole cytoplasm of the germ actively participates in these developmental processes. Though the event in question is here an "external" one, this participation must not on that account be overlooked, for the origin of these external processes must be in the cytoplasm.

From the results of experiments involving merogony and crossing, it is seen that the nucleus, especially at the outset, requires that the cytoplasm should be, as it were, in harmony with it, in order to achieve development. Hence two periods have sometimes been distinguished in early development—the first dominated by the cytoplasm, the second by the nucleus. We need not decide here whether such an antithesis is fully justified; but, in view of the facts, it is impossible to regard the cytoplasm as a passive substratum.

The first cleavage plane of the egg of the newt does not always assume the same direction and position. Sometimes it coincides with the presumptive median plane of the embryo— in which case, if the two ½-blastomeres are artificially separated by constriction, twins are produced. If the cleavage plane is in another direction, a whole germ is sometimes developed from the one blastomere, and only a partial one, without axial organs, from the other. The constriction can be made before cleavage has begun, and it can be so arranged that one half of the egg receives the whole nucleus. If the constriction chance to coincide with the presumptive median plane, it does not matter which half contains the nucleus—a whole germ is formed in any case. If, however, the plane of constriction does not coincide with the median plane, the one half of the egg may produce only a fragment, even though it contain the whole cleavage nucleus. Thus, in spite of the great power of regulation of the egg of the newt, the nucleus—which according to the theory of the gene contains the whole rudiment-material —cannot achieve development. It thus depends entirely upon the cytoplasm of the egg how the developmental fate of the half egg is determined.

In this example we see not only that there is a definite activity of the cytoplasm, but that there proceeds from it something of a determining nature. But more striking proofs of this can be advanced.

It is well-known that cleavage of the Insect egg occurs by the nucleus first undergoing multiple division within the substance of the egg, to produce the so-called cleavage nuclei. These nuclei then gradually pass to the superficial layer of cytoplasm, certain regions of which are taken into their sphere of action. The cleavage cells thus arise, lying superficially, and forming collectively the blastoderm, which covers the whole surface of the egg. Now, there are definite observations which prove that the fate of individual cleavage nuclei, and hence the developmental fate of the regions of the egg, are determined by the superficial layer of cytoplasm. If, for example, a definite region of the surface of the egg of *Musca domestica* be removed before the nuclei have reached the periphery of the egg, they are never able to produce the embryonic structures proper to this cytoplasmic region; and this although the nuclei sometimes wander into the place of operation and multiply there, thus proving that they themselves are not damaged. Even when the nuclei remain longer in the interior of the egg, they are unable to produce these differentiations. Without the peripheral layer of cytoplasm typical development does not take place. Even though it be assumed that differences exist among the nuclei before they enter this layer, yet it is clear that the differentiation of the individual regions of the egg can only be effected by the joint action of the nuclei and the cytoplasm. If this be so, it cannot at the same time be true that the nucleus is exclusively responsible for determination—some part must be played by the cytoplasm.

We have already described how the egg of *Camponotus* shews a very precocious differentiation of its superficial layer of cytoplasm—a differentiation that consists in the laying down of definitely determined germinal regions, as is proved by the later behaviour of the germ (Pl. IV, Figs. 71–74). Here, in this cytoplasmic layer which is predominant in development, we have present already visible differences before the nuclei have reached it; the nuclei do not wander into the various regions of the cytoplasm until later. In this case again it must be assumed that the differentiation of the cytoplasmic layer, which is synonymous with the early development of definite germinal regions, takes place without the aid of the nuclei which later pass into it. In other words, it is the cytoplasm which determines in advance this differentiation.

In this connexion attention may again be drawn to the importance of the yolk-lobe in the egg of the Mollusc, to the pole plasm of the egg of *Tubifex*, and to the cytoplasmic region of the frog's egg called the grey crescent (cf. p. 136). There is no doubt at all that these regions exercise a determining action. Since the differentiations which result from this action consist of typical—i.e. of hereditary—structures, it is at once obvious that the cytoplasm is a repository of genetic factors, even though hitherto it has not been possible to include these factors in the study of heredity. This is due partly to the fact that the characters in question are essential characters, not accessory, and partly to the fact that the method hitherto employed for genetic analysis (crossing) is not applicable to factors of this kind.

The whole process of determination might possibly be conceived, without discrepancy, in terms of a chromosome-mosaic had we to deal solely with typical, normal development; but such a theory breaks down as soon as the processes of regulation are taken into account, and this must be done if we aim at the statement of a general theory of development. This is seen clearly in the experiment already described on the cleavage and development of the egg of the sea-urchin under pressure, in which nuclei pass into cytoplasm which is foreign to them, and yet a normal result is obtained. This can only be explained by making the determination of the germ a function of the state of affairs existing in the substratum, i.e. by supposing that it depends upon the protoplasm as a whole. In the regulation of a young germ whose development has been disturbed, the position of the primary egg-axis relative to the axes of the cleavage cells is also important, thus shewing the influence of the cytoplasmic condition of the egg and the germ upon the essential processes of development.

Without going into greater detail we may conclude, while recognizing the special character of the nucleus as a factor-carrier, that there also exist in the cytoplasm developmental factors which take part in determination. In certain cases, though not in all, this importance of the cytoplasm may be manifested as a visible, morphological egg-structure.

3. The Epigenetic Character of the Genes and the Unity of the Reaction-Basis

From the foregoing considerations there arise two conclusions which may serve as a basis for our further discussion.

The first of these conclusions is one concerning the relative importance of the chromosomes—in particular the constituents of the active nucleus arising from them—in the reaction-basis as a whole. The chromosomes are not the sole "carriers" of determination, they contain only a part of the causes of development, a part which is, indeed, extremely important, but which can become effective only in collaboration with other causes, and in subordination to the system as a whole. Chromosomes are subordinate organizations of the cell which may come to have a relatively great importance of their own, but they alone represent by no means all that is included in the reaction-basis.

The second conclusion in question refers to the individual genes. Directly preforming, corpuscular genes do not exist. On the one hand, their concrete existence is disproved by the general results of analytical embryology, which demonstrate the epigenetic character of all developmental processes; on the other hand, the cytological evidence does not sanction the view that the chromosomes consist simply of discrete individual corpuscles, so that the necessary morphological basis for the genes is lacking.

Experimental analysis of development has further shewn that it consists essentially of an organization which progresses in stages; so that individual steps in organization act as organizers, which then themselves produce new steps in organization. At the beginning of development there is present only relatively little that is organized, and therefore also few organizers, these being of a somewhat general character. (Among these are the chromosomes and the nuclear constituents they produce.) Most structures arise only step by step, and acquire an increasingly specialized character. With regard to developmental factors, this means that those special individual factors that ultimately produce the essential and accessory characters at the end of development are themselves formed only gradually. These "final" factors of development, which have arisen epigenetically, bear a relatively direct relation to the particular characters which "belong" to them. And we now recognize that a synthesis is possible of the basic conceptions of genetics and of analytical embryology, if we bear in mind the epigenetic character of the special factors. These specialized final factors of development are indeed the genes—not existing as directly preforming corpuscles at the beginning of development, but only arising epigenetically as a result of gradual progress in organization during development.

Mendelian analysis leads to the association of individual peculiarities of the phenotype with special factors, which factors have a more or less direct relation to those peculiarities, and form the content of the genotypic formula. These Mendelian factors, or genes, are found by following phenotypic characters through a number of generations—by observation, that is to say, of only the final phase of development. To obtain a correct picture of the primary reaction-basis, the factors so discovered should not be pushed back right to the beginning of development, but be looked for where they become effective—at the end of the process. It has already been pointed out that these Mendelian genes are actually identical with the specialized final factors—a necessary consequence of that epigenetic principle of progressive organization which arises from the experimental analysis of development.

Experimental embryology works up to these factors from below, as it were; while we might say that Mendelian investigation had arrived at them from the opposite direction. But Mendelian analysis itself, by reason of its special method, is unable to penetrate back into the events of development. The region which lies between the beginning of development and these final factors is the domain of analytical embryology, the general conclusions of which go to shew that the final factors, or genes of the Mendelians, arise epigenetically during development. It is not really these genes

which "mendelize," but their indirect basis contained not only
in the chromosomes but in the entirety of the germ-cell, though
chromosomes and other cell-mechanisms may also play their
special parts.

In so far as the genes revealed by Mendelian experiment are
to be regarded as the final factors they and their carriers
have a real existence; though they are by no means simply
discrete corpuscles, and though the necessity for using a
factor-symbol in an inheritance formula furnishes no proof of
the real existence of the gene to which the symbol belongs.
If, however, we elect to pass over a whole period of development, and to transfer the genes to the beginning, they thereby
become merely fictitious entities.

Now, for the purposes of practical genetics, and for the
presentation of a workable theory, this fictive transference of
the final factors—in the form of genes—to the beginning
of development is quite permissible; but we must beware of
imagining that we can get in this way a true picture of the
actual reaction-basis. Since, however, by means of the theory
built upon this fiction we can present most of the known
phenomena of inheritance in an extraordinarily lucid and useful
manner; since, again, this theory enables us to make definite
assertions about the hereditary behaviour of later and earlier
generations, we may say that it completely fulfils its function,
for it performs everything that could be asked of a genetic
theory.

To place the genes exclusively in the chromosomes, and to
assume that these chromosomal genes are the sole decisive
causes of development, is allowable for the special purposes of
genetic theory—if we bear in mind that this is only an
assumption, necessary to the presentation of one particular
theoretical system. The chromosome theory of inheritance
gives, indeed, an extraordinarily clear picture of genetic processes, and for this reason possesses a very great heuristic value.
But it does not present a true picture of what really happens,
nor does it describe the actual nature of the reaction-basis. It
has been constructed on the basis of a study of what are, quite

definitely, partial causes of inheritance and development; so that our notions concerning these causes—the genes—are partly fictitious in nature. In addition to the partial causes of development (supposedly lodged in the chromosomes) which are essential to this theory, there are others which do not lie in the chromosomes. For the purpose of genetics, however, the partial nature of these chromosomal causes can be set aside, and the fiction maintained that they constitute the complete and only cause of development and inheritance, for by doing this the theory fulfils its function. But there exist in the reaction-basis of the whole cell other partial causes of development; and a complete theory of inheritance is conceivable from the standpoint of these non-chromosomal causes alone, though up to the present no such theory has been systematically worked out.

Looked at in this way, a workable theory of heredity, depending on preforming genes and on an exclusively chromosomal basis of development, retains its usefulness when limited to its proper sphere—in spite of facts, derived from experimental embryology, concerning the real nature of the reaction-basis. At the same time, on account of the fictitious ideas involved, and which are necessary for its own purposes, it is ill-adapted to form a criterion of the special nature of development. It is the theory of a particular range of phenomena, and therefore must not be confused with a description of what actually happens; its extension to include a more general theory of development is impossible, chiefly because of its contradiction of the results of experimental embryology. A general theory of development can only come from the investigation of the phenomena of development themselves, not from that statistical distribution of the products of development in the course of generations which, in the main, forms the basis of genetic theory.

The epigenetic conception of the gene limits the chromosome theory of inheritance to its proper sphere, within which it is completely efficient. The chromosome theory provides us with an excellent system for the arrangement, description, and discovery of the facts of the *formal* process of inheritance. Its field, however, is clearly separated from that of experimental embryology. Deductions from it as to the nature of the real reaction-basis are but assumptions, admissible for a definite purpose—the understanding of the hereditary process—and, in the present state of knowledge, perhaps even necessary, but not expressing the actual nature of the reaction-basis.

What is responsible for development is, rather, the whole reproductive cell; it is this in its entirety which we must regard as the reaction-basis; though, as we have said already, there is a definite division of labour within its individuality. It is the specific constitution of the cell that is here the essential thing and which forms the most important physical element in the reaction-basis. Connected with it, and subordinated to it, are a great number of special factor-carriers in the cell. These may be corpuscular (chromosomes) or constitutive, according to whether they are in the nucleus or in the cytoplasm; but in any case they all determine development indirectly rather than directly. Apart from the material factor-carriers, the inter-relations of the factors are essential, and produce further special factors which are non-material.

The chromosomes are important in development and inheritance, but this is not because they are strings of directly preforming, corpuscular genes, but because they are essential subdivisions of the whole cell and collectively furnish certain elements of the active nucleus.

From this standpoint, too, the function of the chromosomes in the formal phenomena of inheritance—segregation, recombination of hereditary characters, etc.—is left untouched. And the linkage of characters is explicable not only by the actual presence of discrete corpuscles in close material union (chromosomes), but might also be easily explained by supposing that the nuclear substances as a whole, represented in the chromosomes, influenced simultaneously the most diverse characters.

Further, we are by no means bound to refer the facts of segregation, etc., solely to a mechanical regrouping of the chromosomes in maturation and fertilization. This is but one of the possibilities that present themselves, though it is the best known. In addition to these morphological and mechanical causes of alternative inheritance, there is the presence of reversible chemical processes, to which great importance is to-day sometimes ascribed. Since we must take into account the constitutive side of the collection of rudiments, the idea naturally suggests itself that rudiment-complexes passing into the zygote from the maternal and paternal sides combine to form a new whole which can be likened to a chemical compound; the segregation of the rudiments can then be attributed to the fact that during the formation of the reproductive cells this compound disintegrates into its constituents, and that new combinations arise from these elements.

Let us, however, leave these objections, directed against the inevitability of a purely morphological theory of heredity, and consider for a moment those subordinate differentiations of the germ-cell which can be called developmental mechanisms. These mechanisms must be regarded as secondary differentiations within the reaction-basis. Just as the developed body has special parts differentiated for special functions, so also the germ-cell has specialized subordinate systems constituting what is secondary, rather than primary, in the reaction-basis. From this point of view, the idea that the general constitution is of capital importance in no way conflicts with a recognition of the role of special morphological subdivisions.

These morphologically recognizable "mechanisms of development," which act as organizers, may not all be present at the beginning of development; they are mostly formed epigenetically during early development—as, for example, in the case of pole plasms, the grey crescent, etc. As regards nuclear substances, however, the facts are rather different. Though these substances must be looked upon as secondary, and subordinate to the whole, they are in a high degree permanent, and are passed from cell to cell—a fact largely responsible for the prominence they have assumed. In the case of other developmental mechanisms this continuity from generation to generation is not so direct. We need only recall in this connexion those organizers which are formed only in the disintegrated germ.

To say that certain material foundations of development in the nucleus—meaning thereby the rudiments contained in the chromosomes—are passed on, as such, from generation to generation, is only true in a restricted sense. For we must remember that the hypothesis of unchanged chromosomes in the active nucleus is not tenable. Thus, even in those parts of the reaction-basis which are borne by the nucleus, there are certain developmental phenomena which are at bottom epigenetic in character, though these may occur so precociously, and bear outwardly so strong a morphological appearance, that they are encountered much more rarely here than in the cytoplasmic parts of the reaction-basis.

The primary foundation of the reaction-basis is not a mosaic of discrete parts. It is the totality of the germ-cell, to which the individual constituents distinguishable in the cell are subordinated—whether these constituents can be passed on to the descendants more or less directly, or whether they must be re-formed at the beginning of each new development, with the whole specific constitution of the reaction-basis as their starting-point. Development is Epigenesis; and not only so as regards the morphological differentiations of the phenotype, but also as regards those subordinate organizations of the reaction-basis which appear only during pre-development. It is epigenetic, too, as regards the "final factors" of development, though the epigenesis may vary in degree. From this standpoint alone, as we have shewn, it is easily possible to reconcile the conclusions of genetics with those of analytical embryology, between which there must otherwise remain a gulf not easy to bridge.

The way is cleared for an evaluation of the organism other than a purely mechanistic one. For the latter arises from the untenable analytical-summative conception of the reaction-basis, and must conflict with the results of analytical embryology when questions of potency and of regulation arise. Thus, however paradoxical it may sound, the study of the "mechanics of development" (*Entwicklungsmechanik*) leads away from the mechanistic conception to a really organismic conception of the organism and therefore of life itself.

BIBLIOGRAPHY

A. BRACHET, L'œuf et les facteurs de l'ontogénèse. Paris, 1931.

J. DUESBERG, L'œuf et ses localisations germinales. Paris, 1926.

B. DÜRKEN, Lehrbuch der Experimentalzoologie. Experimentelle Entwicklungslehre der Tiere, 1928.

E. KORSCHELT, Regeneration und Transplantation. 3 Bände. Berlin, 1927–1931.

T. H. MORGAN, Experimental Embryology. New York, 1927.

E. S. RUSSELL, The Interpretation of Development and Heredity. Oxford, Clarendon Press, 1930.

W. SCHLEIP, Die Determination der Primitiventwicklung. Leipzig, 1929.

E. B. WILSON, The Cell in Development and Heredity. 3rd ed. New York, 1925.

INDEX

(Figures in italics refer to illustrations)